A Study on the Propagation Law of Blasting Seismic Waves in Heterogeneous Media

Huandong Pang

AMERICAN ACADEMIC PRESS

AMERICAN ACADEMIC PRESS

By AMERICAN ACADEMIC PRESS

201 Main Street

Salt Lake City

UT 84111 USA

Email manu@AcademicPress.us

Visit us at http://www.AcademicPress.us

ISBN: 979-8-3370-8955-3

Distributed to the trade by National Book Network Suite 200, 4501 Forbes Boulevard, Lanham, MD 20706

10 9 8 7 6 5 4 3 2 1

PREFACE

Engineering blasting is the process of interaction between explosives and rocks under certain engineering geological conditions by using free surfaces. Therefore, the effectiveness of blasting depends not only on the interaction between explosives and rocks, but also on the engineering geological conditions.

To better understand the process and laws of blasting in the future, we have conducted some calculations and research in an attempt to clarify the changes in the blasting process so as to benefit engineering blasting. The main task is to divide the blasting into different zones based on proportional distance and study their different variation patterns, especially in terms of blasting velocity amplitude and frequency spectrum, which are two crucial factors for the safety of buildings and structures. In addition, the changes in geological conditions of blasting engineering, especially the changes in blasting vibration field in poor inclusions and joints, are discussed and studied, and some verification engineering examples are provided.

A lot of thanks go to my graduate students Tangfei Li, Wei Shen, Chao Pang, Bidan Zhao, Qiyu Pan, for their respective work and contributions to the research of this book.

Given the limits of our knowledge, the time available for research, and

I

the methods we employed, this book is not without shortcomings. We would greatly appreciate it if readers could give constructive criticism and suggestions.

The Author

May 2025

Contents

1 Introduction

1.1 Research Background

Due to the rapid development of China's economic construction and the continuous improvement of people's living standards, the national investment in infrastructure is increasing year by year. Blasting technology, as an important technical means in various construction projects, is widely used in many engineering fields such as water conservancy and hydropower projects, mining, tunnel excavation, and existing building demolition. The wide application of blasting is attributed to its advantages, such as reducing the labor intensity of construction personnel and accelerating the construction progress, which bring great convenience to construction, effectively promote the development of China's infrastructure projects. Blasting technology plays a pivotal role in the rapid construction of the national economy.

With the wide application of blasting technology, blasting operations are increasingly being conducted in urbanized environments and the public awareness of safety and environmental protection is growing. Therefore, the requirements for blasting safety are continuously improved. During the process of blasting, the energy of the explosion is utilized to achieve various engineering purposes. Most of the energy released during blasting is in the

1

form of heat energy, vibration waves, air shock waves. This results in associated hazards like flying stones, blasting vibration, air shock waves, and noise. Among them, blasting vibration, as the primary public hazard, has received widespread attention and concern due to the negative effects caused by blasting operations. How to scientifically and accurately evaluate the damage degree of blasting vibration to surrounding buildings/structures and facilities has always been a central topic in the study of blasting vibration effect.

The propagation of blasting seismic waves in rock media is influenced by many factors, such as the geological conditions and physical and mechanical parameters of the propagation medium, the type and method of blasting, and the size and form of blasting source. The main influencing factors can be divided into two aspects: one is the characteristic parameters of the explosion source itself, such as vibration intensity, frequency and duration; the other is the attributes of the affected objects, such as the geological structures of the blasting engineering and the relative distance from the explosion source. These factors interact and constrain one another. Additionally, if there are unevenly distributed structural planes (such as joints and faults) in the rock mass medium, the refraction and superposition of the waves will occur when the blasting seismic waves propagate to these structural planes, making the propagation of blasting vibration waves more complex. As a result, assessing the damage degree of blasting vibration becomes more challenging.

Under the action of blasting, the failure process of rock media is very complex. When the explosives explode, shock waves generated by high temperatures and high pressures will instantly crush the surrounding rocks. Most of the explosion energy is lost and attenuated into the form of stress waves. As these stress waves propagate outward gradually, the energy attenuates continuously in the process of propagation, and finally propagate in the rock mass in the form of blasting seismic waves. When the charge explodes in the rock, the surrounding rock forms a crushing zone and a fracture zone under the action of explosion. After the stress wave passes through the fracture zone, its strength attenuates rapidly, causing the elastic vibration of rock particles. This kind of vibration propagates outward in the form of elastic waves, resulting in ground vibrations, hence termed as seismic waves. This phenomenon and its consequences caused by blasting are called blasting seismic effect. In the field of blasting engineering, the research on the propagation characteristics of blasting vibration still needs further analysis. How to scientifically evaluate the propagation patterns and changes of explosion vibrations in different media has been extensively explored by engineering technicians and researchers from the aspects of blasting theory, numerical simulation, vibration testing, and stress wave analysis.

Weak interlayers are a geological structure phenomenon existing in naturally deposited rock strata. They are widely present in many large-scale engineering projects at home and abroad, with only less than 70% of the projects not needing to consider the impact of weak interlayers during construction [1]. The presence of weak interlayers is the root cause of many engineering accidents. For example, they led to the instability and failure of the cable crane platform slope on the left bank of the Lancang River Manwan Hydropower Station in Yunnan [2]. They also caused the instability and damage near the dam of the Vajont Hydropower Station Reservoir in Italy, resulting in the death of more than 2600 people and the ultimate failure of the hydropower station's construction [3]. The existence of weak interlayers in the dam foundation and abutment slopes of projects like the Xiangjiaba Hydropower Station caused their anti-slide stability to fail to meet the specification requirements, directly preventing the projects from being completed on schedule [4]. The dam foundations of Gezhouba Water Control Project and Henan Xinyang Chushandian Reservoir and other projects also suffered from poor anti-slide stability due to the existence of weak interlayers, resulting in many problems during construction [5].

When carrying out engineering construction in geological conditions where weak interlays are present, we must pay attention to the harm they may cause. Weak interlayers can have negative impact on the anti-slide stability of rock masses and may even become an important control factor in the construction of dam foundation, chambers and other projects [6]. Particularly under the cyclic dynamic disturbances such as blasting vibration waves, seismic loads, wave loads, weak interlayers are one of the main culprits behind instability disasters in engineering rock masses. At present, the main method to deal with weak interlayers is the replacement method, but not all weak interlayers cause engineering accidents, and using the replacement method consumes significant manpower and material resources, involves substantial engineering work, and lead to considerable waste.

Blasting vibration has typical non-stationary random signal characteristics. The impact of blasting vibration on buildings/structures is essentially a process of energy transmission and transformation, influenced by the amplitude and frequency of blasting vibration, that is, by the energy characteristics of blasting seismic waves. At present, the large-scale construction of foundational engineering projects in China will inevitably lead to the stability and safety problems of surrounding buildings/structures caused by blasting vibration. Therefore, how to ensure the safety of existing buildings/structures as much as possible is still a problem that needs to be studied and solved in engineering construction.

By analyzing the propagation characteristics of blasting vibration velocity and blasting vibration frequency, this study explores the basic law of blasting vibration propagation, which is conducive to better avoiding the economic losses and casualties caused by blasting engineering, so as to formulate more reasonable blasting construction schemes and reduce the impact of blasting vibration propagation.

Secondly, on the basis of mastering the basic theory of blasting vibration, the blasting vibration characteristics at different proportional distances are compared and analyzed. The introduction of proportional distance makes the data more comparable, which is helpful to judge the blasting vibration attenuation law of different blasting schemes in practical engineering, and to evaluate and predict the vibration response of the ground and buildings/structures caused by blasting construction.

1.2 Research at Home and Abroad

The effect of blasting vibration on surrounding buildings/structures is essentially a process of energy propagation and transformation of blasting vibration waves and is affected by the amplitude value and frequency distribution range of blasting vibration waves, that is, by the energy characteristics of blasting vibration waves. To better control the impact of blasting vibration, it is necessary to clarify the characteristics of particle vibration and conduct in-depth research on the propagation characteristics of blasting vibration.

The main parameters of vibration amplitude generated by blasting seismic waves include vibration velocity, acceleration and displacement. There is still no accurate calculation method for the attenuation law of particle vibration velocity amplitude. In practical application, regression analysis based on a large number of measured data is usually carried out to establish the empirical relationship between the amplitude of vibration velocity and various influencing factors. Currently, the research at home and abroad primarily involves adjusting or improving the Sadovsky formula proposed by the former Soviet scholar M. A. Sadovsky based on measured data to adapt to the variation law of blasting vibration in a certain project or a certain range.

Although the research on blasting seismic waves started late in China, a series of rich research results have been achieved through extensive research processes, and corresponding blasting safety regulations have been formulated accordingly. Xie Yushou et al. [7] obtained the propagation characteristics of seismic waves and proposed a formula for the propagation attenuation law of blasting seismic waves by analyzing the blasting vibration monitoring data of different types of rock masses and different geological conditions. Zhang Jietao et al. [8] summarized the attenuation law of blasting vibration velocity at different positions near the blasting area by using regression analysis. Sun Xiuming et al. [9] tested the blasting vibration of the slope on site, obtained the empirical formula for the propagation of blasting vibration, and calculated the safe vibration velocity of buildings at different distances and the safe allowable distance of blasting vibration. He Yunlong [10] obtained the zoning distribution law of blasting vibration acceleration by using the two-dimensional finite element analysis program of blasting vibration, and summarized the approximate calculation method of blasting vibration acceleration generated in the construction of rock slope based on a large number of calculation results. Hu Guozhong et al. [11] studied the ground vibration characteristics under blasting vibration by combining a large number of measured blasting vibration data and seismic wave theory. Zhang Yongzhe [12] obtained the expression of the relationship between the seismic wave velocity, frequency and blasting conditions by analyzing the attenuation

7

law of the seismic wave peak strength, and conducted an in-depth study of the propagation law of the seismic waves in rock masses by analyzing the energy variation characteristics of the seismic waves passing through the fault area. Pang Huandong et al. [13] believed that the seismic wave propagation caused changes in the surrounding media, resulting in changes in the energy composition of the seismic waves. By dividing the action range of the seismic waves into zones, the variation law of the vibration amplitude under different action zones was determined, thereby evaluating the impact effect of the seismic waves on the surface buildings. Wu Delun [14], Tang Chunhai [15] and others developed a new version of *Blasting Safety Regulations* by comprehensively comparing and studying the safety evaluation standards of blasting vibration in different countries and combining with the mature foreign blasting vibration evaluation standards to evaluate the safety and stability of blasting vibration and has been widely used in the field of blasting engineering.

In recent years, some scholars have found that the Sadovsky formula has certain calculation errors in practical engineering applications. Using analysis methods such as the least squares method and empirical formulas, they have conducted preliminary discussions on the optimization and selection of blasting vibration velocity attenuation formulas. Zhang Tianjun et al. [16] studied the blasting vibration characteristics of open-pit mines and found that the Sadovsky formula did not fully consider the influence of elevation amplification effects, resulting in low accuracy and poor reliability in

8

prediction results. Yang Nianhua et al. [17] mentioned that the error in blasting vibration peaks calculated by the Sadovsky formula could reach 200% to 300% at close distances and over 50% at long distances.

Therefore, current research on the propagation law of blasting vibration and safety control mostly involves modifying Sadowsky formula based on the actual engineering projects, studying the attenuation law of blasting vibration amplitude at different positions, and using regression analysis to obtain the appropriate prediction formula. Hao Quanming et al. [18] studied the influence of free surface azimuth on blasting vibration velocity and introduced it into Sadovsky formula for correction. Tao Tiejun et al. [19] established the attenuation formula of blasting seismic wave energy by using the percentage of explosive energy converted into blasting seismic wave energy and the attenuation coefficient of propagation medium, and verified the feasibility of this formula in predicting blasting vibration intensity through engineering examples. Wang Yujie et al. [20] made a comprehensive analysis of the vibration effect of granite under blasting, and used regression analysis method to calculate the attenuation coefficient α and site coefficient K of seismic waves, and summarized the propagation law of seismic waves in intact granite. Through theoretical analysis, Zhang Shaoquan et al. [21] believed that the seismic wave energy conversion coefficient is related to the dielectric coefficient K and attenuation coefficient α in the Sadovsky formula, and gave the theoretical calculation formula of the seismic wave energy conversion coefficient. Tang Hai et al. [22] conducted experimental and theoretical

9

research on the impact of topography and geomorphology on blasting vibration waves, and proposed a vibration formula reflecting elevation.

With the continuous progress of science and technology, the popularity of electronic computers, and the wide application of new monitoring equipment, digital monitoring technology and analysis methods in the field of blasting vibration monitoring, the theoretical research on blasting vibration shows greater room for development. Methods for analyzing the characteristics of blasting vibration include re Fourier analysis, reflection spectrum analysis, wavelet analysis and computer-based numerical simulation. Lou Jianwu et al. [23] used wavelet packet technology to extract the signal characteristics of blasting seismic waves generated after blasting in slightly weathered granite rock field, and found that the waveform of blasting seismic waves can be calculated by the prediction model of wavelet packet coefficient with different frequencies. Ling Tonghua et al. [24] analyzed the attenuation law of blasting seismic wave propagation based on wavelet transform and Fourier transform according to the characteristics of blasting vibration signals, providing an effective analysis technology for the comprehensive study of blasting vibration effects, especially the vibration velocity-frequency related safety criteria. Li Hongtao et al. [25] analyzed the propagation and attenuation mechanism of seismic waves from the perspective of energy, and the results showed that the peak energy and vibration velocity of seismic waves had a double relationship with the attenuation coefficient of distance. In the same blasting, the total energy of blasting earthquake was approximately

10

proportional to the square of particle peak vibration velocity.

In the long-term engineering practice, different industry departments at home and abroad have formulated different safety criteria for different degrees of protection of buildings/structures. The main evaluation indexes are displacement, velocity acceleration and frequency. Due to the increasing complexity of engineering environments, both engineering practice and field monitoring have shown that relying on a single parameter as the safety criterion of blasting vibration is very limited and blasting vibration frequency is also the dominant factor of blasting vibration hazards. Therefore, the peak vibration velocity of particles and the main frequency of blasting vibration are mostly used as the indicators of blasting vibration intensity and safety evaluation. They are used in the analysis and research of blasting vibration, engineering design and safety evaluation.

Zhou Junru et al. [26] used LS-DYNA finite element software to carry out numerical simulation and analysis of the attenuation mechanism and law of blasting vibration frequency of spherical and cylindrical charges. The results showed that due to differences in spectrum characteristics of blasting seismic waves, the main frequency of blasting vibration did not strictly follow a certain law during the attenuation process with the increase of distance from blasting center, and that the attenuation process had certain volatility. However, the average frequency of blasting vibration with the increase of the distance from blasting center had obvious rules to follow. Tang Hai et al. [27] used UDEC program to simulate the variation characteristics of seismic waves

11

in different land forms, and compared them with the measured results of actual projects. The results showed that the amplification effect of land forms on the velocity of seismic waves is not invariable and has a critical point of height. Yu Min et al. [28] conducted on-site monitoring and numerical simulation analysis. Their research results showed that the variation characteristics of particle vibration velocity peak and main frequency were closely related to the height difference. In a certain range, the peak value and main frequency of particle velocity increased with the increase of height difference, but when the height difference exceeds a certain range, the peak value and main frequency of particle vibration velocity decrease gradually with the increase of height difference. Cao Pan et al. [29] simulated and analyzed a series of variation characteristics of explosion stress waves at different distances from the explosion source in rock masses. Ye Haiwang et al. [30] used wavelet packet analysis technology and ANSYS/LS-DYNA numerical simulation software to analyze and study the data measured in the field, and found that the structural plane had obvious weakening effect on the high frequency band of blasting seismic waves. When the vibration wave passes through the structural plane, the peak value of the vibration wave attenuated greatly, with an average attenuation of the vibration velocity of about 40%. Jiang Nan and Zhou Chuanbo [31] deduced the empirical formula of blasting vibration velocity attenuation law considering the influence of the relative slope between the slope measuring point and the blasting source based on dimensional analysis. They established the numerical model of Slope Blasting with different slopes

12

through LS-DYNA for calculation and analysis, and studied the influence of elevation effect on the attenuation of blasting vibration velocity. In order to obtain the attenuation law of blasting vibration frequency and its influencing factors, Lu Wenbo et al. [32] introduced the medium damping term into the solution process of stress waves of spherical charge in elastic media, and established the spectrum expression of blasting vibration in actual rock mass media. Zhou Jianmin [33] used Ansys/LS-DYNA dynamic finite element software to carry out numerical simulation on the Slope Blasting of Zuogou Open Pit Mine, and calculated the velocity, stress, displacement and other dynamic response characteristics of the slope with structural planes under different blasting loads, so as to provide basis for the optimization of on-site blasting parameters and the guarantee of slope stability.

Blasting seismic waves are essentially the energy generated by the explosives during explosion that propagates outward in the form of gravitational waves, and is usually composed of body wave and surface wave. Body waves propagate mainly in the interior of the earth's rock strata, including P-waves and S-waves. The longitudinal wave is a compression wave whose particle vibration direction is parallel to the wave propagation direction. The longitudinal wave propagates outward from the explosion source. Because its vibration direction is parallel to the wave propagation direction, the medium will be compressed and pulled up under the action of the longitudinal waves. Because longitudinal waves propagate along a straight line and their velocity is faster than that of shear waves, it is called the primary wave, characterized by short period, small amplitude and fast propagation. S-wave is the shear wave whose particle vibration direction is perpendicular to the wave propagation direction. Both S-wave and P-wave propagate outward from the explosion source, and the propagation mode directly leads to the shear deformation of the propagation medium. Because S-wave often arrives after the arrival of P-wave, it thus gets the name of secondary wave (S-wave), which is characterized by long period, large amplitude and slow propagation. The surface wave only propagates along the rock interface and the earth's surface. It is formed on the surface of the medium due to the reflection and transmission effects of seismic waves. It is mainly composed of love wave propagating in longitudinal rolling and Rayleigh wave propagating in transverse vibration. It has long period and large amplitude, and its propagation velocity is slower than that of body wave, but it carries large energy [34]. Many scholars at home and abroad have conducted research on the propagation law of blasting seismic waves. As early as the middle of 1920s, Rockwell [35] began to study how the building structure would be affected by the surrounding blasting seismic waves through statistical analysis of a large number of damage phenomena. Later, many scholars began to test the propagation law of blasting seismic waves under different detonation conditions on the basis of his research and studied the various characteristics displayed in the propagation process of blasting seismic waves [36,37]. In 1950, Morris [38] first linked the vibration amplitude of blasting seismic waves with the strength of seismic waves, and, based on this, proposed an

14

empirical formula for the attenuation law of seismic wave amplitude. The quantity of explosives and propagation distance were used as control variables to establish the empirical formula. Carlos [39], Devine [40], Duvall [41] and Dowding [42] studied and modified the empirical formula. By adopting the concept of proportional distance to make the propagation data of blasting seismic waves under different explosive quantities more comparable, they used the peak vibration velocity of particles rather than the maximum amplitude as the basis to determine the strength of seismic waves and whether structures were damaged, making the modified empirical formula recognized by many scholars.

The domestic research on the propagation law of blasting seismic waves mainly began in the middle of the last century. Although it was 30 years later than abroad, with the efforts of many scholars, the blasting safety regulations in line with the world development trend were formulated for China. Zhang Yiping and Wu Guiyi [43] found that the blasting seismic wave is a composite wave generated by the superposition of various waves. Many factors such as the physical and mechanical properties of the propagation medium, the terrain structure, and the quantity of explosives used can affect the propagation parameters and direction of the blasting seismic waves. Zhang Yongzhe [44] obtained the expression of the relationship between the seismic wave velocity and frequency under certain blasting conditions by analyzing the attenuation law of the peak vibration intensity of blasting seismic waves and the different energy changes of blasting seismic waves before and after passing through the fault. Li Aichen [45] carried out blasting vibration tests in the actual slope engineering, obtained an empirical formula of the propagation law of blasting seismic waves in this situation through the engineering measured data and theoretical research, and calculated the safe vibration velocity, critical vibration velocity and safe distance without damage through this formula. MENG et al. [46] believed that blasting seismic waves would affect the properties of surrounding media, thus affecting the composition of blasting seismic wave energy. Therefore, the impact area of blasting seismic waves can be divided under limited conditions. By comparing and analyzing the data of blasting vibration amplitude values in different impact areas, a standard for evaluating the impact of blasting seismic waves on surface buildings can be obtained. On the basis of the established elastic-plastic model, Lin Daneng et al. [47] analyzed the extrusion characteristics of the medium through the cavity forming effect of explosive extrusion, and determined the formula for calculating the extrusion coefficient. Through the calculation and analysis of the parameters of three soil media with different properties and two explosives with different properties, it was found that the influence of the density and strength of the media and the detonation velocity and density of the explosive on the extrusion characteristics of the medium was nonlinear. In order to better describe the attenuation law of seismic waves, Jin Xuhao added a geological quality factor as a control parameter to the traditional empirical formula. Long Yuan et al. [48] proved that the most important factors affecting the propagation law of blasting seismic waves were propagation distance and

16

geological conditions through inductive analysis of a large number of blasting vibration engineering measured data and numerical modeling analysis of blasting vibration. The propagation law of blasting seismic waves was also affected by factors such as the amount of the explosive, the distance from the explosion source, the blasting method, the terrain structure, and the physical and mechanical properties of propagation media. In addition, many scholars [49~54] creatively put forward a series of methods and means to control the propagation of blasting seismic waves to prevent engineering accidents by combining theory with practical engineering. Based on the above research, it can be found that in order to influence the propagation law of blasting seismic waves, the amount of the explosive, the propagation distance of seismic waves, the physical and mechanical properties of propagation media, and the topographic structure should be taken as control variables.

Theoretically, the weak interlayer is very similar to the sandwich biscuit. On both sides of the weak interlayer, there are harder rock masses than the interlayer, but in fact, the weak interlayer can be found in broken faults, structural zones, and fractured rock masses. Thus, weak interlayers do not necessarily exist continuously. Zhang Guojun [55] believes that weak interlayers are essentially structural planes, which have a certain thickness and its strength is far lower than the rock mass covering both sides of the structural plane. Chen Zhixiong [56] and Ren Jie [57] argue that weak interlayers are structurally loose rock layers covered and wrapped by other rock masses, whose mechanical strength is significantly lower than that of the rock masses

17

covered on both sides. They are extremely prone to argillization after water absorption, resulting in rheological effect. Because they are wrapped and covered by rock masses, weak interlayers are of great significance to control the stability of rock masses. Because the physical and mechanical properties or engineering properties of weak interlayers are poorer than those of the rock masses on the left and right or the top and bottom, and because the mechanical strength of some weak interlayers is greatly reduced and argillization occurs after water absorption, weak interlayers, whether they are thick or not, will inevitably cause various obstacles to the engineering construction. When weak interlayers are developed near slopes, landslides, collapses and other accidents are extremely likely to occur under the action of blasting load or rainstorm. Therefore, in order to ensure the safety and stability of slopes, attention should be paid to the impact of weak interlayers [58-59].

Because the physical and mechanical properties of different types of weak interlayers are different, the influence of different types of weak interlayers on the same project is also different. In this case, it is necessary to refine the classification of weak interlayers to facilitate a more detailed study of the impact of weak interlayers on engineering. Many scholars have classified weak interlayers. Wang Guirong [60] studied the material composition of weak interlayers and classified them into the original material composition when the structure is not destroyed, the structural material composition under dynamic action, and the secondary minerals composition under the action of exogenous forces according to the formation causes.

According to particle sizes, weak interlayers can be composed of rock blocks, rock cuttings, coarse particles, fine particles, clay particles, colloidal particles and other components with different proportions. According to the water content, the material composition of weak interlayers can be divided into fluid plastic state material, soft plastic state material, plastic state material and hard plastic state material. According to the three rock categories, weak interlayers can be divided into magmatic rock weak interlayers, sedimentary rock weak interlayers and metamorphic rocks. According to the cause of formation, weak interlayers can be divided into primary weak interlayers, secondary weak interlayers, tectonic weak interlayers, and comprehensive weak interlayers. According to the occurrence of interlayer, they can be divided into horizontal weak interlayers, gently inclined weak interlayers, medium steep weak interlayers, and steep inclined weak interlayers. According to the particle composition and degree of fragmentation, weak interlayers can be divided into argillaceous weak interlayers, broken weak interlayers, and flaky weak interlayers. Hu Tao et al. [61], after consulting a large number of data, classified weak interlayers into three categories according to the engineering classification method: clayey weak interlayers, clay containing silty clastic weak interlayers, and clastic mud mixed weak interlayers, and determined three different correlation formulas for the physical property indexes and strength indexes of the three types of weak interlayers. Li Jingshan et al. [62] also classified weak interlayers into three categories according to their formation causes: the primary weak interlayers of weak rock layers with high clay content, poor cementation and low mechanical strength, the secondary weak interlayers, also known as weathering interlayers, and the structural weak interlayers formed due to tectonic action. They found that if anti-seepage measures were not taken, once local concentrated leakage occurred at weak interlayers under the dam foundation, the possibility of dam damage would be greatly enhanced, and even directly lead to the collapse of dams.

In addition to studies on the classification of weak interlayers, it is also necessary to study the properties of weak interlayers. Research on the properties of weak interlayers began in the 1950s. Ren Guangming [63] believed that the residual strength of some weak interlayers in the process of sliding failure is not the final strength of the weak interlayers. After the sliding failure, the strength of the weak interlayers would rise to a certain extent. Mahr [64] pointed out that weak interlayers in the deep layer under the slope had obvious viscous fluidity and would expand to a certain extent with the passage of time. Tan TK et al. [65] found that one of the main reasons for the long-term deformation of dams is the creep and stress relaxation of weak interlayers in dam foundations under the long-term gravity load. Liu Xiaoli [66] summarized five strength characteristics and corresponding microstructure characteristics of soft soil layer through indoor tests and theoretical analysis, and discussed the existing landslide analysis methods. Jian Wenxing et al. [67] studied the weak interlayer in the Jurassic red bed in Wanzhou. The weak interlayers were thick, continuous and distributed in multiple layers. The study found that the microstructure of the weak interlayers was mainly a sheet

20

structure containing a small amount of grid structure containing a small number of micro cracks and micro pores. Lu Haifeng et al. [68] and Zeng Feng et al. [69] found that the weak interlayers in the red bed area had certain expansibility and their strength was affected by the clay mineral particle composition, the dip angle between the interlayer and the ground, the moisture content of the interlayer, and the undulation and roughness of the contact surface between the interlayer and the upper and lower strata. Huang Qiuxiang et al. [70] believed that the stability of the slope would also be affected by the anti-tilt angle of the weak interlayer. Through the real-time deformation monitoring of the actual project, they found that the overall deformation of the slope was affected by the anti-tilt weak interlayer, which might lead to the failure of the slope due to instability. Zhou Fei et al. [71] conducted a similar test on a large shaking table to study the influence of the thickness of the horizontal weak interlayer on the stability of the slope, and found that the thickness of the weak interlayer had no great impact on the stability of the weak interlayer when other factors were consistent. They found that the slope would be destroyed first as the elevation amplification effect corresponded to the thickness of the weak interlayer. They also summarized the failure mode of the slope with different thickness of weak interlayers. Liu Hanxiang [72] studied the slope with weak interlayers by conducting a large-scale earthquake simulation shaking table test. The results showed that the existence of weak interlayers would enhance the response of PHA and PVA on the slope top, but when the excitation intensity was large, the enhancement effect would be

21

weakened. No matter how the interlayer angle was set, the isolation effect of thick interlayers was stronger than that of thin interlayers, and the weakening effect on component acceleration and component wave in horizontal and vertical directions was better. Song Yanqi [73] conducted a failure test on Fangshan marble. The test results showed that the existence of weak interlayers would weaken the strength of marble, and different tilt angles would weaken the strength of marble differently. It was found that when the dip angle of weak interlayers was 60°, the fracture surface of marble completely coincided with the weak interlayers.

In addition to theoretical analysis and experimental analysis, scholars at home and abroad have also carried out a large number of numerical simulation analyses on weak interlayers. Liu Chuanzheng et al. [74] studied the propagation process of stress waves under dynamic load in weak interlayers by using many theories and found that weak interlayers would lead to multiple reflections and refractions of stress waves at the junction of the interlayer and rock mass, and that the energy of stress waves would be continuously attenuated due to the energy absorption and resistance characteristics of weak interlayers. Yu Jingtao et al. [75] used FLAC3D software to change the inclination angle of the weak interlayer in the model and set five groups of inclination angles of 0°, 15°, 30°, 45° and 60° for comparative analysis. They found that the greater the inclination angle of the weak interlayer was, the worse the stability of the model became and the more likely the tunnel chamber was to be damaged. Sun Jinshan et al. [76-77] studied the propagation characteristics of blasting seismic waves in slopes with weak interlayers by combining theoretical analysis, numerical simulation and similar test methods. The results showed that weak interlayers could significantly weaken the vibration velocity of blasting seismic waves, and that the thicker weak interlayers were, the better the effect would be. A new calculation method of dynamic stress at the sliding surface was given. Huang Feng et al. [78] simulated the process of tunnel excavation near weak interlayers by combining numerical simulation with physical model tests. The results showed that the existence of weak interlayers would make the stress distribution of the tunnel more uneven, leading to the expansion of the tunnel failure zone and ultimately weakening the stability of the tunnel. Du Ruifeng et al. [79] studied the dynamic response characteristics of blasting seismic waves on slopes with weak interlayers through similarity tests and theoretical analyses, and found that weak interlayers could absorb blasting seismic wave to a certain extent, but the hard rock masses above the weak interlayers would amplify the role of blasting seismic waves. Pei Xiangjun et al. [80] and Cui Shenghua et al. [80~82] studied the start-up mechanism of Daguangbao landslide through a series of similar tests and numerical simulations on a vibration test bench. The results showed that the internal stress increased due to the dynamic incompatible deformation between the weak interlayer and the hard rock mass on both sides, ruling out that the sudden reduction of rock mass strength due to fragmentation was not the main reason for the sudden occurrence of landslide, and gave two evolution process of mechanics of the

23

sudden occurrence of Daguangbao Landslide.

Based on the analysis of the above research status at home and abroad, many scholars have conducted in-depth studies on the propagation law of blasting vibration from multiple aspects, such as blasting vibration amplitude or blasting vibration frequency through theoretical analysis, on-site detection and numerical simulation methods. They have made beneficial explorations for the establishment of a comprehensive safety criterion for blasting vibration with multiple factors and have achieved rich research results. However, how to more accurately determine the blasting vibration characteristics at a certain location is still difficult to unify. Therefore, how to comprehensively analyze the propagation characteristics of blasting vibration from both the amplitude and main frequency of blasting vibrations in combination with the changes of engineering geological conditions needs further in-depth research.

1.3 Main Research Contents

Based on the review and analysis of research data at home and abroad and supported by relevant theories and analytical methods such as explosion dynamics, numerical analysis, signal analysis technology, and dynamic finite element theory, this study attempts to use the numerical simulation software ANSYS to study the propagation law of blasting vibrations at different proportional distances. The research results are to be compared and verified with engineering examples. In this study, ANSYS17.0 is used to establish a three-dimensional solid model and carry out pre-processing operations such as meshing, and then LS-DYNA is used to solve the generated K file. LS-PREPOST 4.2 is used for post-processing and Origin8.0 is used for data and image processing. The main research contents are as follows:

(1) Conducting a literature review to get acquainted with the status quo of research on blasting vibration propagation law at home and abroad.

Focusing on the problem of how to better control the impact of blasting vibration on surrounding buildings/structures, this study reviews the current research on the propagation law of blasting vibrations by reading domestic and foreign literatures to look for the breakthrough and innovation of the study.

(2) Carrying out numerical simulations by using conventional models to better reflect the general propagation law.

ANSYS/LS-DYNA numerical software is used for simulations. Based on the elastic wave theory, the element type, real parameters, and material properties are defined to establish a basic model. Operations such as meshing, defining contact information, applying boundary conditions, and loading are performed step by step to carry out simulations.

(3) Extracting data by introducing proportional distance to make the measurement points more comparable.

Based on the distance of measuring points from the explosion center and the quantity of explosives, the corresponding proportional distances are calculated. The correlation coefficient is introduced to divide the explosion area into near, medium and far areas. Combined with the numerical simulation with post-processing software, The blasting vibration velocity data of each measuring point were extracted. The obtained vibration velocity data are processed by using the software MATLAB to perform Fourier transform, completing the transformation from time domain to frequency and thus obtaining the blasting frequency data. The resulting data is then organized and analyzed by using the image analysis software ORIGIN.

(4) Making comparative analyses by combining with the actual engineering projects to verify the rationality of the research results again.

The relationship curves are obtained from the above data. The propagation characteristics of blasting vibration are analyzed from the aspects of the peak value of blasting vibration velocity and the main frequency of blasting vibration, and the propagation laws of blasting vibration under different proportional distances are studied. The research results are compared with and verified by the actual engineering data.

2 Blasting Vibration Law and Variation

After the explosive is detonated, the energy generated by the explosion mainly acts on the surrounding rock masses, causing the rock masses to fracture. The remaining energy disturbs the rock outside the fracture range, generating stress waves and blasting seismic waves in rock masses. The generation and propagation of waves are the result of energy transformation and transmission generated by the blasting of explosives. Under the action of blasting vibration, the dynamic response of the existing buildings/structures is closely related to the propagation characteristics of blasting seismic waves. Once the structures produce vibration response due to the action of blasting, structural damage is bound to occur to a certain extent. In addition, the vibration generated by blasting causes psychological panic to people. In view of the influence of blasting vibration on the existing structures on the surface, it is necessary to understand the propagation characteristics and attenuation law of blasting seismic waves.

2.1 Generation of Blasting Seismic Waves

Blasting vibration waves are shock waves generated by explosion in rock mass medium. According to the nature, waveform and degree of action on rock mass medium, the propagation process of blasting vibration wave can be divided into three action zones, as shown in Figure 2.1. The explosion of the explosives in the sandstone medium forms instantaneous shock waves. The intensity of the waves propagating to the outside world decrease with the increase of distance, and the nature and waveform of the waves also change.

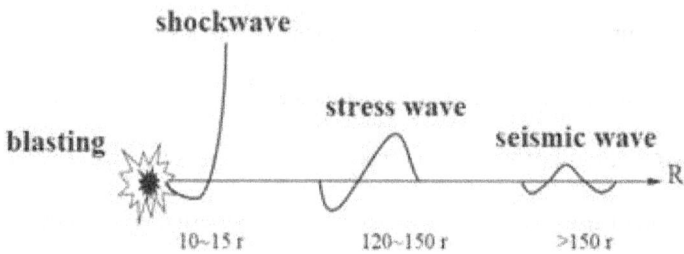

Fig.2.1 Schematic diagram of shock wave, stress wave and seismic wave

Within a short distance of about 10~15 times the radius of the charge from the explosion source, the strength of the blasting vibration shock waves is extremely high and its peak pressure generally exceeds the compressive strength of the rock. Under the action of the shock waves, the rock rapidly produces large plastic deformation. At the same time, most of the impact energy is consumed in the process of shock waves crushing surrounding rocks, and the strength also decreases sharply. This zone is called the shock wave action zone.

After passing through this zone, the shock waves attenuate into stress waves without steep wave crest due to the large amount of energy consumption. The change of state parameters on the wave front is relatively gentle and the wave velocity is close to or equal to the sound velocity in the

28

rock. The time required for the state change of the rock is far less than the time required to return to the static state. Due to the effect of stress waves, the rock is in an inelastic state, producing plastic deformation in the rock, and even leading to failure. This zone is called stress wave action zone or compressive stress wave action zone. Its range can reach 120~150 times the radius of the charge.

When the blasting vibration propagates to a long distance, the strength of the waves further attenuates and becomes elastic waves or seismic waves. When the explosive explodes in the soil and rock medium, only 2%~20% of the energy is converted into seismic waves, and its range is beyond 150 times the radius of the charge. At this time, the propagation velocity of the waves is equal to the velocity of sound in the rock. Its role can only cause the rock particles to make elastic vibration and generally will not cause the rock to break. However, the deformation and displacement of joints and fissures in the rock may become invalid. The time when the rock particles leave the static state is equal to the time when they return to the static state. Thus, this zone is called the elastic vibration zone.

Blasting seismic waves are elastic waves composed of a variety of vibration waves. According to different propagation paths, blasting seismic waves can be divided into body wave and surface wave. Body wave refers to the propagation in the medium and surface wave refers to propagation along the surface of the medium. At the same time, according to the directions of wave propagation and particle motion, the body wave is divided into P-wave and S-wave, and the surface wave can be divided into Rayleigh wave and love wave as shown in Figure 2.2.

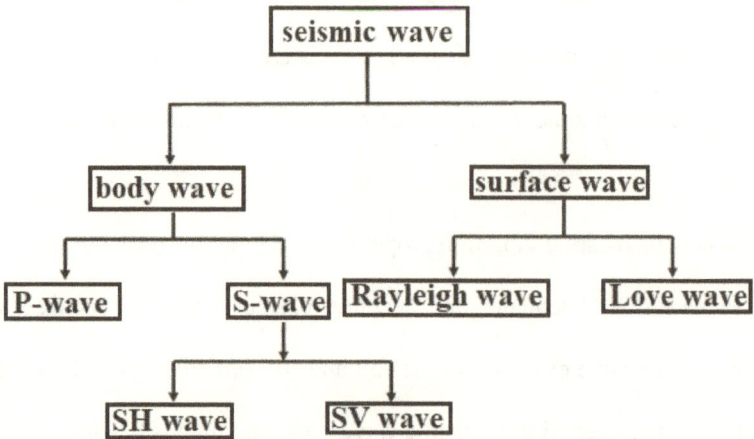

Fig.2.2 Classification of blasting seismic waves

P-wave is one type of body wave. Its particle displacement direction is consistent with the propagation direction of the wave. When it propagates outward from the source, it can cause compression and tensile deformation of the medium. P-wave has the characteristics of short period, small amplitude and rapid propagation. The commonly-seen types of P-wave are compression wave, irrotational wave and density wave. Among them, the P-wave that produces expansion effect without rotation is called irrotational wave. In the process of propagation, the P-wave that causes the medium particles to show the stretching and compression deformation of reciprocating motion is also called compression wave.

Transverse wave is also a kind of body wave. It can cause the shear deformation of particles in the propagation medium, thus it is also called shear wave. It has the characteristics of long period, large amplitude and slower propagation velocity than longitudinal wave. According to the position of particle motion direction and vibration propagation direction, shear wave can be divided into SH wave and SV wave. The vibration direction of SH wave is parallel to the propagation direction, whereas the vibration direction of SV wave is perpendicular to the propagation direction, as shown in Figure 2.3.

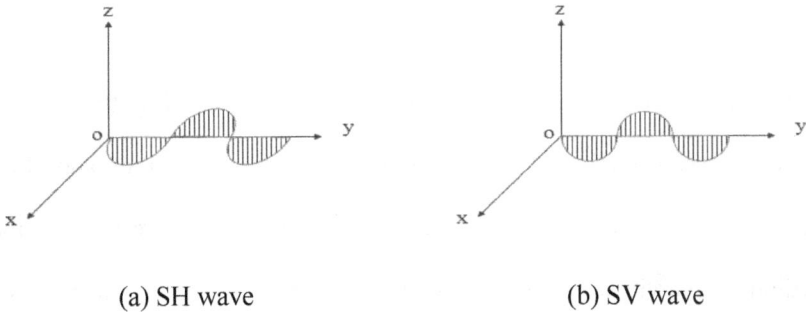

(a) SH wave   (b) SV wave

Fig.2.3 Two modes of propagation of shear wave

Surface wave is formed by superposition of body waves after multiple reflections on the free surface. It has the characteristics of long period, large amplitude, slowest propagation velocity, and large amounts of energy. Due to the different trajectories of particles, surface wave is further divided into Rayleigh wave and love wave. Rayleigh wave is formed by the interference of longitudinal wave and SV wave near the free surface of particle medium, and its motion direction is parallel to the propagation direction and is elliptic. Love wave is similar to shear wave, which propagates along the layer in layered media, and its motion direction is perpendicular to the wave

31

propagation direction.

2.2 Wave Equation of Blasting Earthquake

As a kind of elastic wave, the propagation process of blasting seismic waves is a traveling disturbance and a reflection of energy transfer from one point of rock mass medium to another as well. The blasting seismic force applied to the elastic body of soil and rock cannot be immediately transmitted to all parts of soil and rock within the blasting area, but is gradually transmitted outward in the form of elastic wave through the deformation caused by the blasting seismic force. Therefore, the propagation of elastic zone in rock and other media obeys the law of elastic wave propagation. In the wave process, the medium is in a vibrating state. Assuming that the particle in the medium is a unit hexahedron, its motion equation can be derived from Newton's second law, namely Navier's equation:

$$
\begin{aligned}
\rho \frac{\partial^2 u}{\partial t^2} &= \frac{\partial \sigma_{xx}}{\partial x} + \frac{\partial \sigma_{yx}}{\partial y} + \frac{\partial \sigma_{zx}}{\partial z} \\
\rho \frac{\partial^2 v}{\partial t^2} &= \frac{\partial \sigma_{xy}}{\partial x} + \frac{\partial \sigma_{yy}}{\partial y} + \frac{\partial \sigma_{zy}}{\partial z} \\
\rho \frac{\partial^2 w}{\partial t^2} &= \frac{\partial \sigma_{xz}}{\partial x} + \frac{\partial \sigma_{yz}}{\partial y} + \frac{\partial \sigma_{zz}}{\partial z}
\end{aligned}
\tag{2.1}
$$

In the above formula, ρ is the unit volume density, σ_{xx}, σ_{yx}, σ_{zx}, σ_{xy}, σ_{yy}, σ_{zy}, σ_{xz}, σ_{yz}, σ_{zz} are the stress components of the particle element hexahedron respectively, and u, v, w are the displacement components of the element hexahedron along the x, y, z directions.

If displacement components u, v, w are used to represent the wave equation, it is:

$$(\lambda + G)\frac{\partial e}{\partial x} + G\nabla^2 u - \rho\frac{\partial^2 u}{\partial t^2} = 0$$

$$(\lambda + G)\frac{\partial e}{\partial y} + G\nabla^2 v - \rho\frac{\partial^2 v}{\partial t^2} = 0 \qquad (2.2)$$

$$(\lambda + G)\frac{\partial e}{\partial z} + G\nabla^2 w - \rho\frac{\partial^2 w}{\partial t^2} = 0$$

Of which, e is the unit volume strain; $e = \varepsilon_x + \varepsilon_y + \varepsilon_z$; λ is lame coefficient; G is shear modulus; and ∇^2 is Laplace operator.

In the infinite elastic body, the wave equation of blasting seismic wave longitudinal wave (P wave) and transverse wave (S wave) can be derived from equation (2.2).

In equation (2.2), adding the X differential in the first equation, the Y differential in the second equation, and the Z differential in the first equation in turn can obtain the general wave equation of the longitudinal wave (P wave) as follows:

$$\frac{\partial^2 e}{\partial t^2} = C_p^2 \nabla^2 e \qquad (2.3)$$

Of which, C_p is the longitudinal wave velocity; $C_p = \sqrt{\frac{\lambda + 2G}{\rho}} = \sqrt{\frac{E(1-v)}{\rho(1+v)(1-2v)}}$; v is the Poisson's ratio of rock mass.

In equation (2.2), the Z-differential in the second equation and the Y-differential in the first equation are successively subtracted to obtain the general wave equation of shear wave (S-wave) as follows:

33

$$\frac{1}{2}\frac{\partial^2}{\partial t^2}\left(\frac{\partial w}{\partial y}-\frac{\partial v}{\partial z}\right)=\frac{1}{2}\frac{G}{\rho}\left(\frac{\partial^2}{\partial x^2}+\frac{\partial^2}{\partial y^2}+\frac{\partial^2}{\partial z^2}\right)\left(\frac{\partial w}{\partial y}-\frac{\partial v}{\partial z}\right) \quad (2.4)$$

Let $\varpi_x=\frac{1}{2}\left(\frac{\partial w}{\partial y}-\frac{\partial v}{\partial z}\right)$, $C_s^2=\frac{G}{\rho}$, deform equation (2.4) to obtain:

$$\frac{\partial^2 \varpi_x}{\partial_t^2}=C_s^2 \nabla^2 \varpi_x \quad (2.5)$$

Of which, C_s is shear wave velocity; $C_s=\sqrt{\frac{G}{\rho}}=\sqrt{\frac{E}{2\rho(1+v)}}$; ϖ_x is the angular displacement around the x axis.

The above are the wave equations of S-wave and P-wave in the x axis direction, and the wave equations in the y and z directions can also be derived as above.

The process of blasting seismic waves propagating outward is actually the end stage of the process in which explosion energy expands, does work, is consumed and attenuates in the medium. When the explosive detonates, it releases a large amount of energy, which causes a sudden increase in temperature and pressure inside the rock mass and produce a large number of gases. These gases expand at a very fast speed and impact the nearby rock mass and form a shock wave. The propagation velocity of this shock wave in the rock mass is far faster than that of the blasting shock wave, and the dynamic compressive bearing capacity of the rock is often far lower than the strength of the blasting shock wave. Therefore, the rock mass within the radius of the explosive charge of 3~7 times is crushed by the shock wave and forms a crushing area. Because the shock wave consumes a lot of energy when it breaks the rock mass or causes large deformation of the rock mass, and the energy consumed in the hard rock with large breaking resistance is more intense, the energy level of the shock wave decreases rapidly. When the shock wave crosses a distance of 120 times the radius of the charge, it attenuates into a stress wave. Such a stress wave loses the ability to break the rock mass, but causes the rock mass to produce a displacement away from the direction of the explosion source. This displacement causes a tensile stress in the rock mass that exceeds the tensile strength of the rock mass, resulting in the generation and expansion of cracks in the rock mass. At the same time, a large amount of gas generated by the release of a large amount of energy with the detonation of explosives enters the cracks and causes the cracks to expand and extend. When the stress wave in the rock mass is eliminated, the energy accumulated by the compression of the rock mass is released, and the shear stress that deviates from the propagation direction of the stress wave is generated. This shear stress forms circumferential cracks in the rock mass, and a large number of radial and circumferential cracks pass through, resulting in the destruction of the rock mass. When the propagation distance of the stress wave exceeds 150 times the radius of the charge, the energy level further decreases, which is called seismic waves. Because the energy level of the seismic waves is far less than that of the shock wave and stress wave, the rock mass can only vibrate within the linear elastic range, and this elastic vibration continuously attenuates to zero with the increase of the propagation distance. This zone is called the elastic vibration zone.

As the distance from the center of the explosion source increases, a large amount of energy released by the explosion of the explosive is continuously dissipated. When the energy propagates over a distance of 120 times the radius of the charge, the shock wave energy level decreases and transforms into a stress wave. When the energy propagates over a distance of 150 times the radius of the charge, the stress wave further attenuates in energy level and becomes an elastic seismic wave that propagates outward. The vibration of the medium particles near the explosive under the action of the energy generated by blasting is called the seismic wave, and the sum of the vibration of the relevant medium particles caused by the outward propagation of the seismic wave in the medium is called the blasting earthquake [83]. The essence of blasting vibration is the process that the seismic wave generated by the explosion of the explosive causes the adjacent particles in the medium to move back and forth in a straight line or curve along their equilibrium position. The formation process of blasting seismic wave is shown in Figure 2.1.

2.3 Propagation Law and Characteristics of Blasting Seismic Waves

Explosive blasting produces huge energy, most of which is consumed in breaking the surrounding rock mass and causing large-scale deformation of the surrounding rock mass. The remaining small portion of energy spreads outward in the form of stress waves and makes the medium particles move, thus affecting the safety of surrounding buildings and structures. The propagation of blasting seismic waves are affected by many factors, which can be mainly divided into the following three aspects:

(1) Explosion source parameters. Factors such as the properties of explosives, burial depth, blasting mode, maximum charge volume per section, charge structure and other parameters affect the attenuation velocity, vibration frequency and vibration amplitude of energy in the process of blasting seismic wave propagation.

(2) Propagation media. When the blasting source parameters and geological structure are fixed, the properties of the propagation medium make the blasting seismic wave energy levels different. Most of the energy generated by the explosion is not converted into blasting seismic waves. In dry soil, only 2% to 3% of the energy is converted into blasting seismic waves; in wet soil, only 5% to 6% of the energy is converted into blasting seismic waves; in rock, only 2% to 6% of the energy is converted into blasting seismic waves; and in water, about 20% of the energy is converted into blasting seismic waves. From the above conversion efficiency, it can be seen that the more uniform the medium is converted into blasting seismic waves, the more explosion energy will be, and the higher the blasting vibration frequency and amplitude will be. In short, the propagation medium greatly affect the consumption of blasting energy and vibration effect.

(3) Geological structure. Geological formations, which are formed over billions of years under a variety of actions, are complex and diverse, with rare repetitions. Rock layers often contain a large number of fractures, faults and

other discontinuities rather than being uniform and continuous. When seismic waves encounter these structural planes, they continue to produce reflection and refraction effects, constantly altering their propagation paths and directions. Therefore, the propagation path and process of seismic waves change with the variations of geological structure.

Blasting seismic waves exhibit the following key propagation characteristics:

First, complexity. The diversity of blasting source parameters, the diversity of topographic conditions, the complexity of geological structure and the difference in physical and mechanical properties of media determine that blasting seismic waves are complex and difficult to replicate.

Second, instantaneity. Blasting is a process that occurs instantaneously. The propagation time of blasting seismic waves is extremely short, and the blasting energy attenuates rapidly.

Third, diversity. Different topographic conditions, different physical and mechanical properties of different media, and different structural planes of different geological structures can cause various changes in the propagation process of blasting seismic waves. The reflected and refracted seismic waves superimpose and combine with the newly arriving waves, making he blasting seismic waves highly diverse.

2.4 Reflection and Transmission of Blasting Seismic Waves

The propagation of blasting seismic waves is a complex problem of wave mechanics. The propagation of blasting seismic waves in media has the characteristics of energy property, randomness, complexity and diversity.

(1) Energy property

Blasting seismic waves carry certain energy in the process of propagation, including the kinetic energy of particle motion and the elastic strain energy of particles. Part of the energy released by explosion is converted into blasting seismic waves, which transmit energy outward through particle motion. As the propagation distance increases, the wave front expands, and the wave energy attenuates due to absorption by the medium's damping effect. The energy attenuation of blasting seismic waves is related to the medium's damping and wave frequency. High-frequency waves are more susceptible to damping and absorption by the medium, thus attenuating slowly in the medium.

(2) Complexity

The generation and propagation of blasting seismic waves involve a complex process influenced by many factors. Firstly, the seismic waves occur after the explosion of explosives and they are affected by factors such as explosive performance, charge quantity, borehole arrangement, charge structure, and detonation modes, leading to variations. Secondly, wave propagation is also related to rock properties, and different rock properties lead to great differences in the wave propagation process. In the process of propagation, when seismic waves propagate to the interface of heterogeneous media, different forms of waves are reflected and transmitted. The superposition and interference of these waves are also the main factors leading to the complexity of wave propagation.

(3) Diversity

A variety of blasting seismic waves with different characteristics will be generated in the process of blasting, including body wave, surface wave, direct wave, refraction wave and reflected wave. They have different propagation velocity and propagation characteristics in the medium. In the process of propagation, they are superimposed to produce stress waves with different phases and frequencies. The waveform parameters change continuously with time, reflecting the diversity of blasting seismic wave propagation.

(4) Randomness

Due to the complexity and variability of geological structure and rock properties, blasting seismic waves exhibit randomness in the process of propagation. The randomness of blasting seismic waves is also reflected in the continuous change of stress waves with time. Even when blasting is monitored at the same place and under the same conditions, the same waveforms do not appear, showing certain differences.

It is assumed that when an incident seismic wave I reaches the interface in the process of propagating from rock medium I to joint medium II, both rock medium I and joint medium II are disturbed. The generated reflected

wave r returns to rock medium I for propagation, and the generated transmitted

wave T enters joint medium II for propagation through the interface. The

whole process is basically continuous without separating the two media for

propagation, as shown in Figure 2.2.

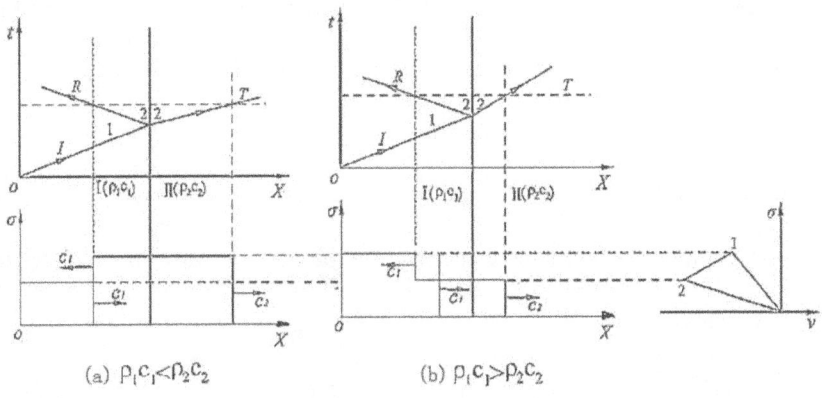

(a) $\rho_1 c_1 < \rho_2 c_2$ (b) $\rho_1 c_1 > \rho_2 c_2$

Fig. 2.2 Reflection of seismic waves on the interface of joint medium

Due to the continuity of stress wave propagation and the law of

conservation of energy, the particle velocity and stress on both sides of the

interface are equal:

$$\frac{\sigma_T}{\rho_2 c_2} = \frac{\sigma_I}{\rho_1 c_1} = -\frac{\sigma_R}{\rho_1 c_1} \qquad (2.6)$$

$$v_R + v_I = v_T \qquad (2.7)$$

$$\sigma_R + \sigma_I = \sigma_T \qquad (2.8)$$

Simultaneous (2.6) - (2.8) can be obtained:

$$\frac{\sigma_R}{\sigma_I} = \frac{\rho_2 c_2 - \rho_1 c_1}{\rho_2 c_2 + \rho_1 c_1} \qquad (2.9)$$

$$\frac{\sigma_T}{\sigma_I} = \frac{2\rho_2 c_2}{\rho_2 c_2 + \rho_1 c_1} \tag{2.10}$$

$$\frac{v_R}{v_I} = \frac{\rho_2 c_2 - \rho_1 c_1}{\rho_2 c_2 + \rho_1 c_1} \tag{2.11}$$

$$\frac{v_T}{v_I} = \frac{2\rho_2 c_2}{\rho_2 c_2 + \rho_1 c_1} \tag{2.12}$$

Obviously, according to formula (2.9), formula (2.10), formula (2.11) and formula (2.12), the different wave impedances of rock and joint lead to different properties of reflection and transmission of stress waves, and the incident and transmission directions of waves are consistent, while the size and direction of reflected waves depend on the ratio of wave impedances of the two media.

(1) When $\rho_1 c_1 < \rho_2 c_2$, there are $\sigma_R > 0, v_R > 0, v_T < v_I$. It shows that the reflected wave R and the incident wave I have the same sign, and the particle vibration directions of the two media are the same.

(2) When $\rho_1 c_1 > \rho_2 c_2$, there are $\sigma_R > 0, v_R < 0, v_T < v_I$. It shows that the reflected wave R and the incident wave I have different signs. When the compressive stress wave is incident, the reflected tensile stress will be generated at the interface between the two media.

(3) When $\rho_2 c_2 \to \infty$, there are $\sigma_T = 2\sigma_I, v_T = 0$. It shows that the incident wave I is reflected by the rigid wall at the interface between rock and joint and cannot enter the joint medium for propagation. At this time, the particle vibration velocity of joint medium II is 0, and the stress of joint

medium II is exactly twice the incident stress.

(4) When $\rho_1 c_1 = \rho_2 c_2$, there are $\sigma_R = 0, v_R = 0, v_T = v_I$. It shows that there is no reflected wave r at the interface of the two media, and the incident wave I is completely transmitted into the jointed medium II, which is equivalent to propagating in the same medium.

(5) When $\rho_2 c_2 \to 0$, there are $\sigma_T = 0, v_T = 2v_I$. It shows that the incident wave I is completely reflected, and the interface between the two media is a free surface. At this time, the compression wave and the tension wave reflect each other.

3 Numerical Simulation of Rock Mass Blasting

3.1 Division of Blasting Areas

To ensure the safety of surrounding structures under blasting vibration, the ground vibration intensity should be controlled within the critical value specified in *Safety Regulations for Blasting*. Exceeding this critical value may cause damage to the protected objects. Therefore, controlling the blasting vibration intensity is the key to managing the blasting hazards. The intensity of blasting vibration is related to many parameters. It is generally believed that the safety of a location mainly depends on its distance from the explosion source, that is, the distance from the explosion center. Therefore, in the process of rock mass numerical simulation, monitoring points should be carefully selected and data analysis should be carried out.

Different blasting positions have different effects on the near and far areas, and the laws followed are also very different. To make the data of measuring points more comparable, the concept of proportional distance is introduced. Measuring points are set in the horizontal radial direction of the explosion source, and the blasting vibration simulation results at different measuring points are extracted to more truly reflect the vibration characteristics in the actual blasting engineering.

In general, an explosive with a radius a in an infinite medium is detonated, and then the seismic waves propagate around. Its propagation law satisfies the wave equation. When the stress wave in the center of the charge propagates uniformly, if Φ is taken as the displacement scalar potential function, there is

$$\frac{\partial^2 \phi}{\partial t^2} = V_P^2 \left(\frac{\partial^2 \phi}{\partial R^2} + \frac{2}{R} \frac{\partial \phi}{\partial R} \right) \tag{3.1}$$

Here V_P is the longitudinal wave velocity; R is the distance from the explosion center; and $R \geq a > 0$, $t = \tau - \frac{R-a}{V_P} \geq 0$.

Initial condition:

$$\Phi(R,0) = \dot{\Phi}(R,0) = 0 \tag{3.2}$$

After blasting $t = 0$, the pressure in the explosion chamber changes exponentially:

$$p = P_0 e^{-bi}, \quad b > 0 \tag{3.3}$$

In equation (3.3) the displacement solution of the particle has the following form:

$$u = A'R^{-2} + B'R^{-1} \tag{3.4}$$

Of which A' and B' are weight parameters, which depend on the rock mass, geological conditions and wave propagation time. After comparison, it can be found that the particle displacement is composed of two parts. In the area close to the explosion source, $\frac{1}{R} \ll \frac{1}{R^2}$, the particle displacement is mainly

45

determined by the first item of equation (3.4); in areas far from the source, $\frac{1}{R} \gg \frac{1}{R^2}$, the particle displacement is mainly determined by the second item of equation (3.4).

The specific values of A' and B' in equation (3.4) are generally difficult to obtain. To obtain the numerical solution in the general sense and to facilitate the comparison of different scales of blasting, the blasting center distance is converted into the proportional distance for processing. Therefore, the definition of proportional distance is shown in equation (3.5):

$$\bar{R} = \frac{R}{\sqrt[3]{Q}} \qquad (3.5)$$

Wherein, R is the distance from explosion centers and Q is the maximum amount of explosive charge.

3.2 Numerical Simulation Method

Finite element method is a powerful numerical calculation tool to solve actual engineering problems. It is a numerical analysis method that organically combines elastic theory, computational mathematics and computer software. LS-DYNA is a fully functional program for geometric nonlinearity, material nonlinearity and contact nonlinearity. Mainly adopting Lagrange algorithm, along with ALE and Euler algorithms, it focuses on explicit solving and structural analysis, but still with the capabilities of implicit solving, thermal analysis and fluid structure coupling. It is mainly based on nonlinear dynamic analysis, but it also has the function of static analysis (such as prestress calculation before dynamic analysis and spring back calculation after sheet stamping). Due to its good nonlinear processing ability and powerful fluid structure coupling calculation, LS-DYNA can effectively solve the practical engineering problems of various types of explosions (underwater, in the air, in surrounding rock masses).

ANSYS/LS-DYNA is a well-known explicit dynamic analysis software with powerful and accurate finite element simulation performance. With a large number of different types of element models, material models and algorithm choices, it can easily handle various highly nonlinear problems, such as various collision analysis, stamping analysis, explosion analysis, drop analysis, thermal analysis, fluid solid coupling analysis, etc. It is widely used in automotive, defense and military industry, aerospace, electronics, petroleum, manufacturing and construction industries.

In terms of pre-processing and post-processing, LS-DYNA adopted FEMB of ETA company in the early stage, and developed the post-processing program LS-POST. After further research and development, LS-PREPOST1.0 was developed on the basis of LS-POST post processor. The program has both post-processing function and certain pre-processing function. With the continuous improvement and perfection of the pre-processing and post-processing functions, LS-PREPOST has become a very powerful LS-DYNA pre-processing and post-processing software. It is widely used because of its advantages such as advanced visualization of post-processing, animation demonstration of cloud data, d3plot-oriented historical information rendering of user defined data.

In its later development, LS-DYNA cooperated with ANSYS and jointly launched the ANSYS/LS-DYNA software, which combines ANSYS interface's powerful pre-processing and post-processing, unified database and LS-DYNA solver's powerful nonlinear analysis, making ANSYS/LS-DYNA the most ideal choice for solving various highly nonlinear transient problems.

ANSYS/LS-DYNA uses the simulation analysis environment of ANSYS, and its program organization is similar to other analysis modules of ANSYS, that is, the program architecture is divided into two layers: Begin Level and Processor Level.

Begin Level is the level where the user enters and leaves the ANSYS/LS-DYNA analysis environment. At this level, the user can perform the following operations:

(1) enter the related processors of Processor Level and switch between processors;

(2) clear all the data of the current work to start a new work;

(3) change the name of the current work.

In the GUI environment, the above functions are not limited to Begin Level, but there are still restrictions in the process of batch command flow. For example, only through Begin Level, can the switch from one processor to another be completed. Processor Level is composed of a series of processor programs and LS-DYNA solver programs. In different programming stages, different solvers need to be used for corresponding operations to complete the whole analysis process. To enter and exit the processor or switch from one processor to another processor, one can directly click the stand-alone menu items in the GUI. The corresponding commands in batch mode are shown in Table 3.1.

Table 3.1 Command to enter or exit a processor program

Operation command	function
/prep7	Enter the pre-processor PREP7 program of ANSYS
/solu	Enter LS-DYNA solver program
/post1	Enter ANSYS general postprocessor POST1
/post26	Post processor post26 after entering ANSYS time history
finish	Any processor returns to the start layer, and then switches to another processor through the start layer

Similar to the operation process of general CAE auxiliary analysis programs (such as the structural analysis module of ANSYS), a complete explicit dynamic analysis process of ANSYS/LS-DYNA includes three basic operation links: pre-processing, loading solution, and post-processing. ANSYS is used for the pre-processing of modeling; then LS-DYNA is used for numerical calculation; and finally LS-PREPOST is used for post-processing analysis of the calculation results as shown in Figure 3.1.

In the pre-processing with ANSYS, the K file is generated by defining the parameters such as element properties, materials, contact types, loads, boundary conditions, and solution time. The generated K file is modified to specify keywords according to the model requirements and then imported into ANSYS/LS-DYNA for solution. After the solution calculation, a d3plot file is generated. The post-processing link is then begun. That is, LS-PREPOST is used to open d3plot to get the required feature analysis model animation. The analyses of stress, strain, displacement and time-history curve are carried out. To sum up, the whole pre-process of and post-processing and solving is organically organized as shown in the above figure.

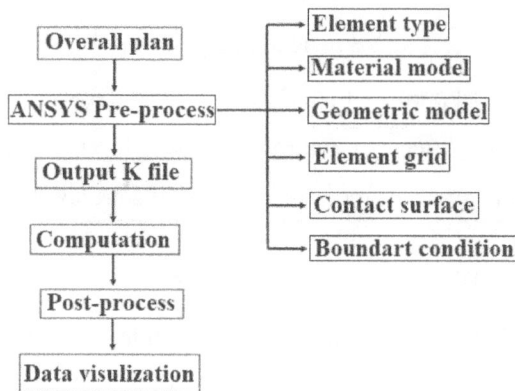

Fig.3.1 Flow chart of numerical calculation of ANSYS/LS-DYNA

In the simulation of blasting test, the numerical analysis algorithm is the top priority of the numerical process. The explosion usually occurs in an instant, accompanied by a strong stress wave. The equation of state and the material model of explosives in LS-DYNA can accurately simulate the instantaneous corresponding process when blasting occurs. Therefore, after establishing the finite element calculation geometric model, it is necessary to select appropriate algorithms, so as to effectively save the calculation time. There are many algorithms provided in ANSYS/LS-DYNA, including the three basic algorithms of Lagrange algorithm, ALE algorithm, and Euler algorithm. The three algorithms can be controlled by ELFORM in the keyword *SECTION_SOLID. Different values of ELFORM lead to different element algorithms. The element formulas available for fluid structure interaction analysis are shown in Table 3.2:

Table 3.2 Value of element formula for fluid structure interaction analysis

ELFORM value	unit algorithm
1	Constant stress solid element (pure Lagrangian algorithm)
5	ALE algorithm for central single point integration (single material in cell)
6	Eulerisn element with central single point integration (single material in the element)
7	Environmental Eulerisn element for central single point integration (used for inlet and outlet boundary of Eulerisn calculation)
11	Ale multi-material unit with central single point integration (a unit can contain multiple substances)
12	Single material ale element with blank material by central single point integration

(1) Lagrange algorithm

Lagrange algorithm is mainly used to analyze the stress and strain problems in solid structures. In this algorithm, the coordinate system and the numerical calculation model are connected, and the material points and element nodes are also integrated. The finite element point is the material point. The basic principle is that the vertices of each mesh move with the filling materials in the mesh, and the filling materials always remain inside the original cells and will not move between cells. The application of Lagrange algorithm can well describe the deformation of materials and make the boundary of material structure deformation very clear.

Lagrange algorithm is often used in the simulation of solid materials and small deformation problems. Its advantages include accurately capturing the boundary interface of the material structure, fast program operation, and low computational cost. However, because its grid division is based on material coordinates, substances cannot flow between the unit grids, and the divided grid move and deform together with the material under study. Therefore, with significant limitations, the Lagrange algorithm is more suitable for small deformation problems of materials. When the material structure deformation is too large, the grids deform greatly along with the material, resulting in serious distortion of the element grid. Therefore, the numerical calculation is difficult to continue, even leading to the collapse of calculations. At the same time, as the minimum element size controls the time step, the calculation time will be doubled if the divided grids are too small, thus affecting the efficiency of calculation.

(2) Euler algorithm

Euler algorithm is mainly used to analyze fluid motion problems and highly material nonlinear problems such as large displacement and large deformation. The algorithm is based on the spatial coordinates, and the divided meshes and the analyzed material structure are independent of each other. The initial spatial position of the grid remains unchanged throughout the analysis process. The finite element node is a spatial point, and the location of the space remains unchanged throughout the analysis process. That is, the material point and the finite element point are not a unified point, but can be separated. The coordinate system is independent of the spatial coordinate system of the calculation model. The Euler grid does not deform with the movement of the material, and the position and shape of the grid are fixed. Due to this characteristic of the grid, it is very difficult for Euler algorithm to track the boundary interface of material structure, resulting in low calculation accuracy and high calculation cost.

(3) ALE algorithm

ALE (algebraic Lagrange and Euler) algorithm was initially applied to the field of fluid dynamics, but it has the advantages of both Lagrange algorithm and Euler algorithm. Firstly, on the basis of Lagrange algorithm, it increases the flexibility of the grid. Its internal meshes can be arbitrarily specified, and the element meshes and material can move separately, which can effectively track the boundary movement of the material structure. Secondly, the Euler algorithm is continued in the division of the internal meshes. The motion of the element grid and the motion of the material

53

structure are separated from each other. The position of the grid can be adjusted appropriately in the process of solving according to the defined parameters so that the grid will not be seriously distorted and the deformation rate of the grid will be reduced. This makes ALE algorithm very advantageous in analyzing large deformation problems. Nowadays, it is also widely used in the simulation of fluid solid coupling problems. Therefore, ALE algorithm is used to calculate and analyze the blasting vibration in this study.

ALE algorithm first performs one or more Lagrange time steps, during which the element grids deform as materials flow. It then performs ALE time step calculations. Step 1: Maintain the boundary conditions of the deformed object and re-divide the internal element grids and keep the topological relationship of the grid unchanged, which is known as Smooth Step. Step 2: Transfer the element variables (such as density, energy, and stress tensor) and node velocity vector in the deformed grid to the new meshes after re-division, which is known as Advance Step.

The user can select the start and end time of ALE time step and its frequency. Euler algorithm is that the material flows in a fixed grid. In LS-DYNA, the relevant solid elements are marked with Euler algorithm and the transportation algorithm is selected. LS-DYNA can also easily couple Euler grid with Lagrange finite element meshes, and has dealt with the interaction between fluid and structure under various responsible load conditions.

3.3 Numerical Simulation of Uniform Material Blasting

The size of the model affects the overall shape and stability of the structure. Therefore, the size of the rock mass model should be reasonably selected according to the actual blasting cases to ensure that the analysis results are more scientific and convincing. Before establishing the model, the following assumptions and simplifications were made to the model according to the experimental requirements:

(1) It is assumed that the rock mass is homogeneous and isotropic;

(2) It is assumed that the rock mass is the same kind of rock mass with stable rock mass property;

(3) It is assumed that the mechanical parameters of the rock mass under blasting are similar to the static load mechanical parameters measured in the laboratory.

Based on an engineering example, this study uses ANSYS/LS-DYNA numerical software to analyze the changes of rock blasting vibration at different proportional distances under the blasting action of 4kg explosive. In the modeling process, in order to simplify the calculation process, a 1/4 rectangular rock mass model with a model size of 80m × 20m × 60m is selected. The overall model unit size is appropriate. The specific model establishment is shown in Figure 3.2.

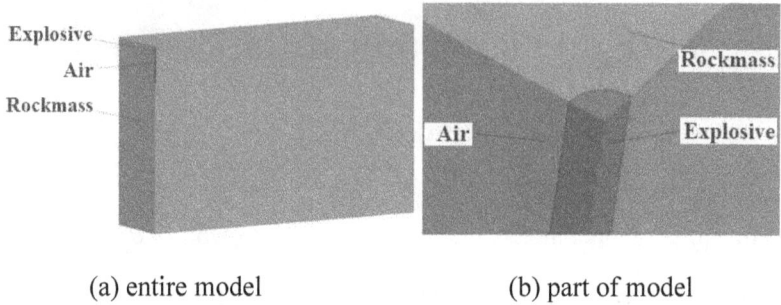

(a) entire model   (b) part of model

Fig.3.2 Establishment of numerical model

To ensure the consistency of calculation parameter units, the basic unit of model calculation parameters is set as cm-g-μS. In mesh generation, three parts are defined respectively: the rock mass material is part 1 and the Lagrange grid is used; the air is part 2 and the explosive is part 3, both of which use Euler grids. Additionally, in the numerical software, the elements of 2D solid 162 and 3D solid 164 are the most suitable for blasting vibration analysis. Therefore, this study selects the 8-node 3D solid 164 element for solid modeling and establishes a total of 407,316 elements.

For the selection of mesh size, if the mesh size is smaller, the calculation accuracy of the model will be higher, but the calculation time will be increased accordingly. To avoid waveform distortion and ensure the accuracy of calculation, the element size must be approximately less than one tenth to one eighth of the wavelength. In this modeling, the smaller-sized elements are distributed around the blastholes, with the sizes ranging from 0.1m to 0.3m, while the rock mass elements have sizes of 0.5m~1.5m. The mesh division of the model is reasonable in density and the specific division is shown in Figure

3.3.

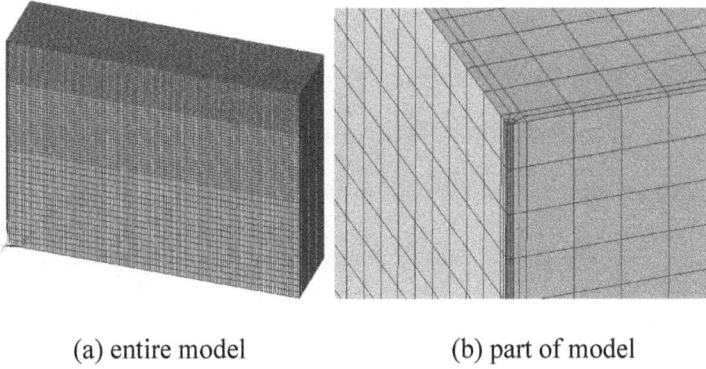

(a) entire model   (b) part of model

Fig.3.3 Mesh generation

There are many factors that affect the propagation of blasting vibration. In order to avoid the influence of other factors on blasting vibration, this study discusses the propagation characteristics of seismic wave vibration in general, and uses the * Holmquist Johnson cook (HJC) model to simulate rock mass materials. This model can better describe the nonlinear deformation of engineering and geological materials in large strain and high strain rate, and is widely used in the impact explosion of engineering materials. The HJC constitutive model describes the relationship between yield stress σ_y, damage and strain rate $\dot{\varepsilon}$ as follows:

$$\sigma^* = \begin{cases} A[\dfrac{p^*}{T^*} + (1-D)](1 + C \ln \dot{\varepsilon}^*) & p^* < 0 \\ [A(1-D) + Bp^{*N}](1 + C \ln \dot{\varepsilon}^*) \le SMAX & p^* > 0 \end{cases} \quad (3.6)$$

Here $\sigma^* = \sqrt{3J_2}/f_c'$, J_2 is the second invariant of deviatoric stress tenso, f_c' is uniaxial dimensionless compressive strength; $p^* = p/f_c'$ is dimensionless pressure; $T^* = f_1/f_c'$ is dimensionless uniaxial tensile strength; $\dot{\varepsilon}^* = \dot{\varepsilon}/\dot{\varepsilon}_0$ is dimensionless strain rate ($\dot{\varepsilon} = 1.0$ s^{-1} is the reference strain rate) rate; A, B, N and C are constants, which can be determined by the material test of rock specimen; Smax is the dimensional maximum strength. At the same time, the HJC model takes into account the effects of shear damage and compression damage, and the damage parameter D can be expressed as:

$$D = \sum \left[\frac{d\varepsilon_p + d\mu_p}{\varepsilon_p^f + \mu_p^f} \right] \qquad (3.7)$$

$$\varepsilon_p^f + \mu_p^f = D_1(p^* + T^*)^{D_2} \geq EF_{MIN} \qquad (3.8)$$

Of which $d\varepsilon_p$ is equivalent plastic strain increment; $d\mu_p$ is irreversible volumetric strain-increment; $(\varepsilon_p^f + \mu_p^f)$ is the plastic strain of rock mass material at fracture; D_1, D_2 and EF_{MIN} are parameters.

At the same time, the physical and mechanical parameters of granite are selected, and the material parameters are shown in Table 3.3

Table 3.3 Calculation parameters of rock mass model

$\rho/(g/cm^3)$	E/GPa	A	B	C	N	FC	T	EPS0	EFMIN
2.1	14.8	0.79	1.6	0.007	0.61	48	4	0.001	0.01
SFMAX	PC/MPa	UC	PL/MPa	UL	D1	D2	K1/GPa	K2/GPa	K3/GPa
7	16	0.001	800	0.1	0.04	1	85	~171	208

Due to some technical defects of spherical explosive charges in the process of blasting construction, cylindrical explosive charges are used in this simulation for their advantages of more uniform explosion energy distribution in rock masses, high energy utilization rate, and better rock crushing quality.

In LS-DYNA, the high explosive model *MAT_HIGH_EXPLOSIVE_BURN is often used to simulate explosive materials, and the relationship between pressure and volume during the explosion of explosives is calculated by combining JWL equation of state to describe the pressure of detonation products. The JWL equation of the explosive with density ρ_0 contains six parameters A, B, R_1, R_2, ω, E_0 (C in isentropic form). When the ideal explosive explodes, the detonation products satisfy the constrained conservation equations in the CJ state. Therefore, before determining the six parameters of the JWL equation of state, the detonation velocity D, detonation pressure P_{CJ} (or isentropic index γ) and detonation heat Q of the explosive need to be known. The expression is as follows:

$$P = A\left(1 - \frac{\omega}{R_1 V}\right)e^{-R_1 V} + B\left(1 - \frac{\omega}{R_2 V}\right)e^{-R_2 V} + \frac{\omega E_0}{V} \tag{3.9}$$

Where P is the pressure of explosive detonation products; V is the initial relative volume; E_0 is the initial specific internal energy of explosive per unit volume; A, B, R_1, R_2 and ω are the material constants describing the JWL equation, and R_1, R_2 are the quantities of dimension one.

Table 3.4 Parameters of charge for calculation

ρ (g/cm³)	D (cm/μS)	A /GPa	B	R_1	R_2	ω
1.2	0.500	2.14	0.0182	4.15	0.95	0.38

In the numerical simulation of explosion shock, air is usually simplified as an inviscid ideal gas. Therefore, the air is described by *MAT_NULL material model. To define the ideal gas material model, we need to define its state equation, which has linear polynomials, i.e. *EOS_LINEAR_POLYNOMIAL and Gruneisen. The equation of state Gruneisen is an adiabatic and entropy increasing equation of state. In this study, the linear polynomial equation is used as the equation of state of air, in which the pressure and internal energy of air have a linear relationship as follows:

$$P = c_0 + c_1 V + c_2 V^2 + c_3 V^3 + (c_4 + c_5 V + c_6 V^2)e \qquad (3.10)$$

Where P is the material pressure; e is the internal energy density per unit volume; V is the relative volume; and c_1, c_2, c_3, c_4, c_5, c_6 are polynomial coefficients. Specific parameters are shown in Table 3.5:

Table 3.5 Calculation parameters of air model

Density $\rho/$ (g/cm³)	c_0	c_1	c_2	c_3	c_4	c_5	c_6
0.00129	0	0	0	0	0.4	0.4	0

In the process of numerical simulation, reasonable selection of boundary conditions can help to improve the effectiveness of calculation. In actual engineering, blasting vibration wave is attenuated in an infinite space, but in numerical simulation, infinite space cannot be established. Therefore, it is necessary to set boundary conditions to simulate infinite space with a finite model.

When establishing the finite element model, this study adopts 1/4 of the model to impose symmetrical constraints on the two vertical sides adjacent to the explosive. The surface is set as a free surface without boundary treatment, and non-reflective boundaries are imposed on other directions to simulate infinite rock masses, which more truly reflects the transmission of blasting vibration waves in rock masses and achieves the coupling effect.

Before the solution is formally started, to ensure the smooth progress of numerical simulation calculation and accurately simulate the site working conditions, the parameters and options related to the solution and result output are set, including solution time, solution step size, CPU control time, solution termination criteria, energy control parameters, hourglass control coefficient, and volume viscosity coefficient.

This simulation defines the solution time as 60MS, the solution step size as the default value, the CPU control time as the default value, and the energy setting coefficient as the default value. the corresponding K file is generated and the pre-processing of ANSYS is completed.

After modifying the *MATERIAL, *SECTION, *CONTROL OPTIONS and other parts of the K file, the LS-PREPOST post-processing software is used to observe the whole blasting process and obtain the equivalent stress nephogram of rock blasting within the calculation time, as shown in Figure 3.4.

It can be seen from Figure 3.4 (a) that after the explosive is detonated, a cylindrical stress wave surface centered on the charge is rapidly formed, with obvious detonation pressure. The equivalent stress continues to increase and reaches the maximum impact stress within 0.9MS after the detonation. The rock mass within 3m away from the explosion source is damaged, as shown in Figure 3.5.

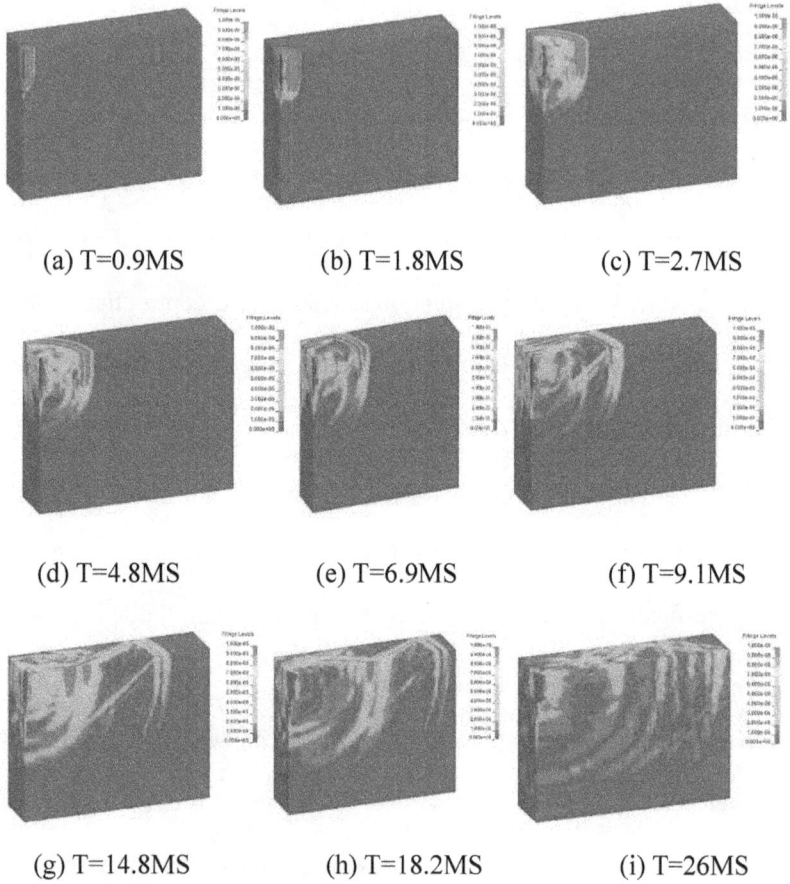

| (a) T=0.9MS | (b) T=1.8MS | (c) T=2.7MS |

| (d) T=4.8MS | (e) T=6.9MS | (f) T=9.1MS |

| (g) T=14.8MS | (h) T=18.2MS | (i) T=26MS |

Fig.3.4 Equivalent force nephogram of rock mass explosion

As the shock wave propagates outward from the center of the explosion source, the stress wave pressure decreases, and the red part in the stress nephogram gradually spreads outward. When t=1.8MS, the stress wave range increases and the color of the stress cloud around the explosion source changes from red to green, indicating that the equivalent stress begins to attenuate when the maximum stress value is reached during the propagation process.

LS-DYNA user input
Time =   908.69
Contours of Effective Stress (v-m)
min=0, at elem# 2730
max=0.000428952, at elem# 102861

Fig.3.5 Schematic diagram of local rock mass failure at 0.9MS of

blasting action

When t=0 to t=4.8MS, the propagation velocity of blasting vibration is faster and the equivalent stress of blasting vibration is in a state of rapid and stable attenuation, which can be seen from the above cloud chart. When t=6.9MS, due to boundary constraints, the stress shock wave begins to reflect and generate epigenetic waves after contacting the rock boundary. According to the equivalent stress nephogram of t=9.1MS, t=14.8MS and t=18.2MS, the shock wave generated by blasting vibration continues to spread outward. At the same time, the reflected blast wave also propagates outward in the rock mass medium, resulting in the superposition of blasting vibration waves. When t=26MS, there is basically no red part in the blasting nephogram, and the seismic wave attenuation is elastic stress wave. According to the above cloud chart, it can be concluded that the propagation process of vibration waves is in good agreement with the objective laws, and the calculation results are reliable.

Based on the above calculations and analysis, the following conclusions can be drawn:

(1) This chapter studies the generation and characteristics of blasting vibration waves and describes the classification of blasting seismic waves and the different propagation modes and attenuation laws of different types of waves on the surface. It also introduces the program organization form, numerical simulation process and solution algorithm of ANSYS/LS-DYNA software.

(2) In this study, ALE algorithm is selected to calculate the fluid-solid coupling process of cylindrical explosive charge in rock mass blasting when ANSYS is used for numerical simulation. A 1/4 model with a size of 80m × 20m × 60m is established, with a total of 407,316 elements and reasonably divided meshes. At the same time, the *MAT\U HIGH\U EXPLOSIVE\U BURN high-energy explosive model is used to simulate explosive materials, and the rock mass uses the *Holmquist-Johnson-Cook (HJC) model. The air is described by *MAT_NULL material model.

(3) LS-PREPOST post-processing software is used to obtain the blasting

equivalent force nephogram. After the explosion of the explosive, the maximum vibration shock wave is reached within 0.9MS, and the rock mass 3m away from the explosion source is in a damaged state. During the time period from t=0 to t=4.8MS, the blasting vibration shock wave gradually propagates outward, and its stress value shows a trend of attenuation according to the change of cloud image color. However, after 6.9MS of blasting action, during the attenuation process of blasting vibration waves, epigenetic waves are generated due to boundary constraints, and the vibration waves are superimposed. When t=26MS, the blasting vibration wave attenuates to the state of elastic stress wave.

4 Analysis of Numerical Simulation Results

4.1 Vibration Velocity of Measuring Points

According to the model and formula (3.5) established in the previous chapter, eight measuring points are selected along the horizontal radial direction the explosive with the proportional distances of $1\text{m/kg}^{1/3}$, $5\text{m/kg}^{1/3}$, $10\text{m/kg}^{1/3}$, $15\text{m/kg}^{1/3}$, $20\text{m/kg}^{1/3}$, $30\text{m/kg}^{1/3}$, $40\text{m/kg}^{1/3}$, $50\text{m/kg}^{1/3}$ respectively as shown in Figure 4.1. By analyzing the simulation results of monitoring points #1~#8, the propagation law of blasting vibration at different proportional distances under blasting action is studied.

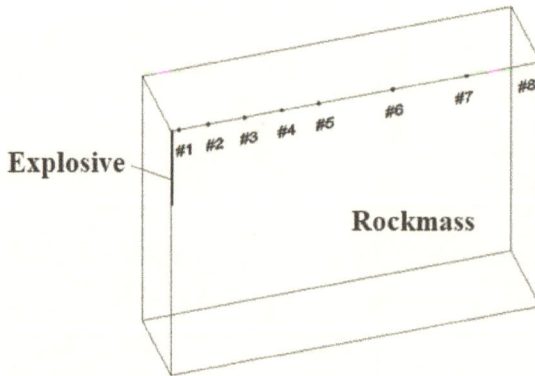

Fig.4.1 Location map of rock mass measuring points

The velocity time-history curves of different measuring points with time as the horizontal axis and vibration velocity as the vertical axis were extracted, and the vibration propagation law in the blasting process was analyzed and summarized. The study of blasting seismic waves only focuses on the attenuation process of blasting vibration amplitude without considering the refraction and emission of seismic waves because the blasting vibration waves have different degrees of refraction and reflection in actual engineering.

There is a good correlation between blasting vibration intensity and particle vibration velocity, which can better reflect the stress at the particle of blasting seismic waves. Through the analysis of particle peak vibration velocity, the relationship between structural internal force and vibration velocity can be established. At present, Sadovsky formula is the most widely used, which is shown in formula (4.1):

$$v = K \left(\frac{Q^\beta}{R} \right)^\alpha \tag{4.1}$$

Where v is the ground particle vibration velocity, cm/S; Q is the single charge, kg; K and α are the coefficients related to the propagation medium; β is the coefficient related to the charge structure; 1/3 for concentrated charge and 1/2 for cylindrical charge. In recent years, more and more scholars have found that the calculation results deviate from the actual value due to the fact that Sadovsky formula does not consider the effect of elevation difference. Therefore, in engineering practice, the blasting parameters should be adjusted according to the specific blasting scheme and geological conditions, and the numerical simulation software should be used to predict so as to ensure the accuracy of calculation.

4.2 Analysis at Closer Points

Based on the establishment of the model, the simulated data of measuring points #1~#4 are extracted from the post-processing software, and the velocity time-history curves in the X, Y and Z directions are sorted out by using the image software Origin. The propagation law of vibration velocity under blasting is explored when the proportional distance $\bar{R}$=1, $\bar{R}$=5, $\bar{R}$=10, and $\bar{R}$=15.

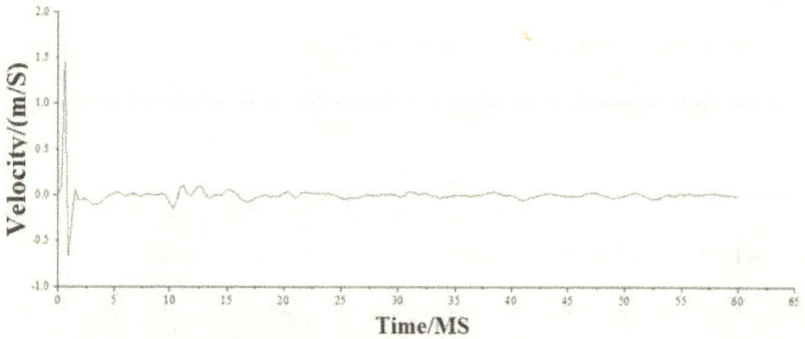

Fig.4.2 Velocity time-history curve in the X direction at position 1

According to Figures. 4.2~ 4.6, the velocity time-history curves in the X direction at the positions where the proportional distance $\bar{R}$ is $1\,\mathrm{m/kg^{1/3}}$, $5\,\mathrm{m/kg^{1/3}}$, $10\,\mathrm{m/kg^{1/3}}$ and $15\,\mathrm{m/kg^{1/3}}$ (For brevity, the following text maybe omits the unit) are analyzed and the following observations can be obtained.

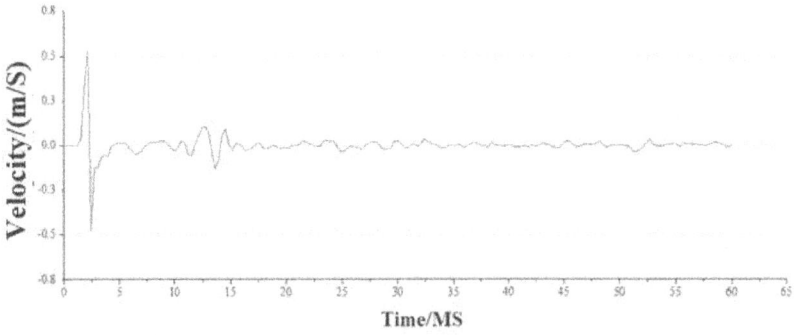

Fig.4.3 Velocity time-history curve in the X direction at position 2

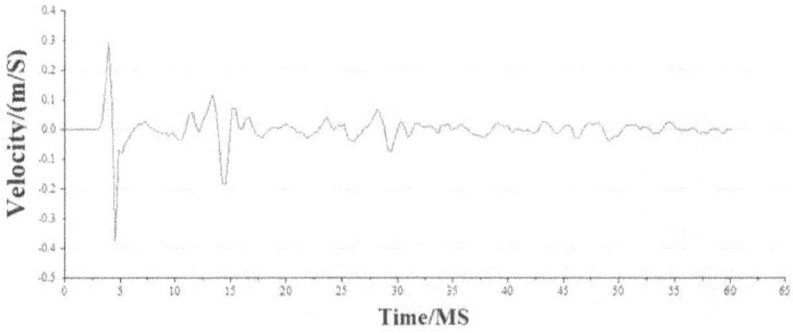

Fig.4.4 Velocity time-history curve in the X direction at position 3

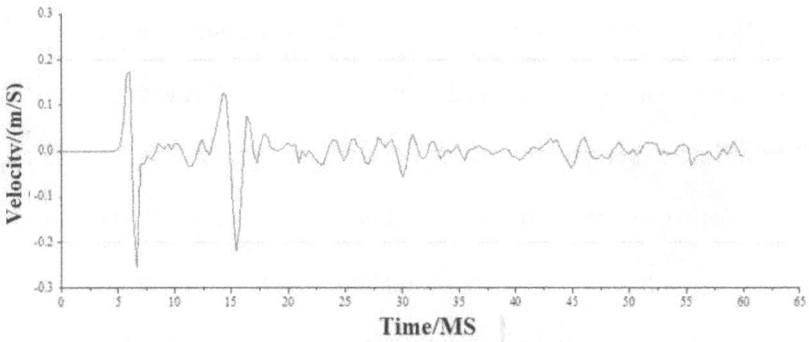

Fig.4.5 Velocity time-history curve in the X direction at position 4

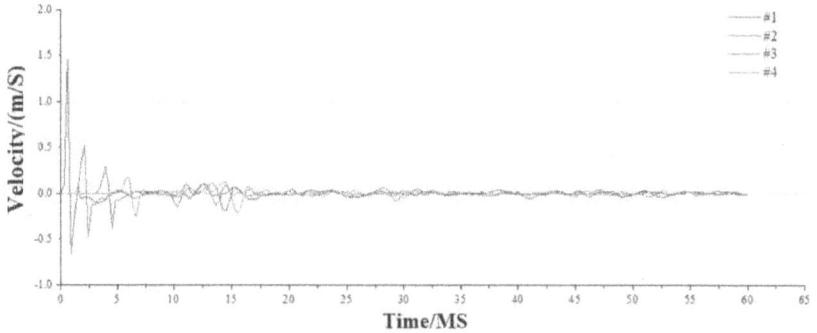

Fig.4.6 Velocity time-history curves under the action of blasting in the X direction at position #1 to #4

(1) After the explosive is detonated, the vibration velocity at measuring point #1 fluctuates instantaneously. The maximum value of blasting vibration velocity in the X direction during the whole blasting process is obtained after 0.61 MS of blasting, which is 1.456224 m/S, and its minimum vibration velocity is obtained after 0.91m/S, which is 0.659565m/S. After 2.5 MS, the fluctuation amplitude decreases rapidly and its vibration velocity tends to be gentle.

(2) The maximum vibration amplitude at measuring point #2 showed a sudden downward trend compared with that at position #1, with a decrease of 63.0%. The maximum vibration velocity was 0.5296737m/S at 2.12 MS after the detonation, and the minimum vibration velocity was 0.4798412m/S after 2.42 MS. In addition to the blasting vibration within 0~5 MS, measuring point #2 also exhibits minor velocity fluctuations during the 10~15MS period.

(3) Position #3 is in a static state within 2.5MS after the detonation of explosives. At 3.94 MS under the blasting action, the maximum vibration

70

velocity in the X direction is 0.2935324m/S, which is 45.3% lower than the peak value of blasting vibration velocity at measuring point #2. At 4.54MS, the minimum vibration velocity is reached, which is -0.3772656 m/S. Compared with the waveform at measuring point #2, the vibration velocity also fluctuates in 25MS~30 MS, and the periods of small fluctuations increase.

(4) At position #4, the vibration velocity remains static during the period of 0~5MS. The maximum vibration velocity is 0.1742073m/S at 6.06 MS, and the minimum vibration velocity was 0.2549558 m/S at 6.66MS after detonation. The peak value of blasting vibration velocity decreases by 29.6% compared with that at position #3. The blasting vibration velocity shows a large fluctuation in the period of 10MS~20MS again, and after 20MS, the fluctuation tends to be stable and continues to oscillate until 60MS.

By comparing and analyzing the time-history curves of blasting vibration velocity in the X direction at measuring points #1~#4, it can be seen that the waveforms are roughly similar. The peak vibration velocity of particles is declining and the time to reach the peak vibration velocity of particles is gradually delayed. The maximum value of vibration velocity between measuring points decreases from 63% to 29.6%, and small fluctuations occur again within 10~20MS after the detonation of explosives.

As shown in Figures 4.7~4.11, from the velocity time-history curves in the Y direction at position #1~#4 under blasting action, the following analysis can be obtained.

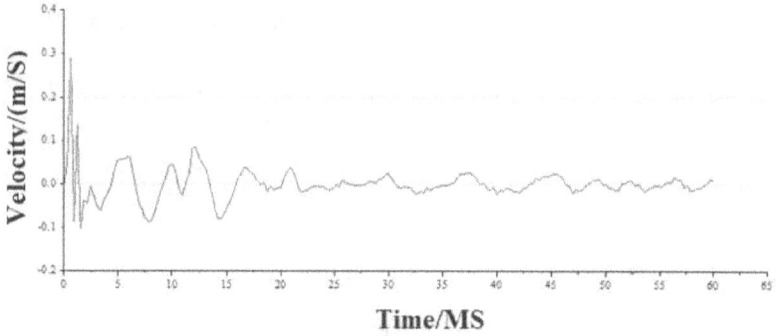

Fig.4.7 Velocity curve in the Y direction at position 1, $\bar{R} = 1$

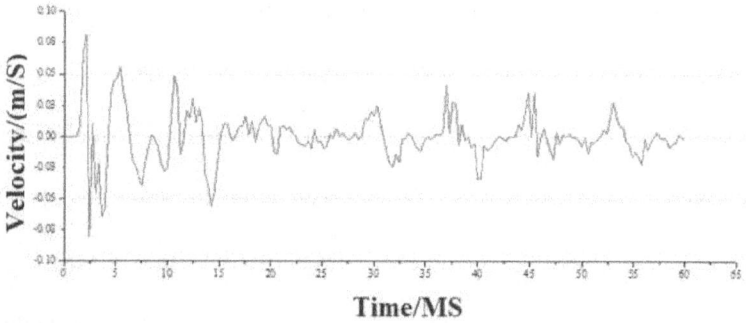

Fig.4.8 Velocity curve in the Y direction at position 1, $\bar{R} = 5$

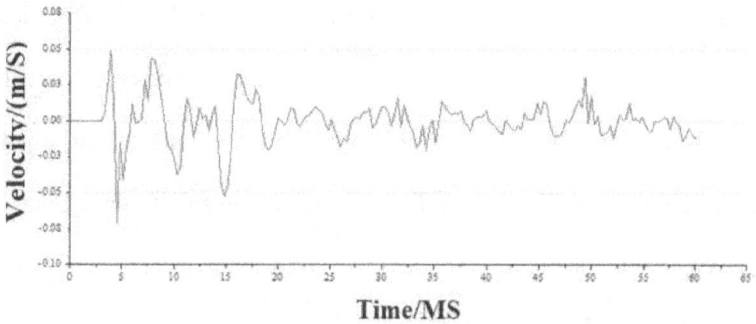

Fig.4.9 Velocity curve in the Y direction at position 3, $\bar{R} = 10$

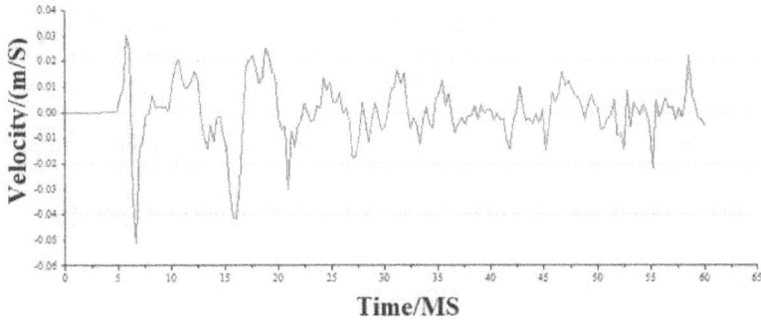

Fig.4.10 Velocity curve in the Y direction at position 4, $\bar{R} = 15$

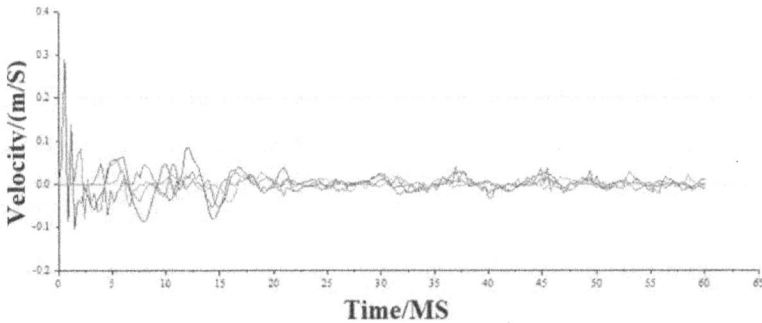

Fig.4.11 Velocity time-history curves in the Y direction at

positions #1 to #4

(1) Under the action of 4kg explosive, the velocity time-history curve is similar to that in the X direction at the same position. Strong vibration is produced after the detonation of the explosive, and within 2.5MS, the maximum vibration amplitude is reached, with the maximum and minimum vibration velocities being 0.2894094m/S and -0.1019706m/S respectively. The difference from the velocity time-history curve in the X direction is that after the maximum amplitude is reached, obvious continuous fluctuation in vibration velocity can be observed in the time period of 2.5MS~17.5MS, and

the fluctuation tends to stabilize and produces oscillations after 20MS.

(2) The vibration velocity fluctuates after 1MS at position #2 and reaches the maximum vibration amplitude at 2.12MS of blasting action. The maximum vibration velocity is 0.08208225m/S, decreasing by 72.4% compared with the peak vibration velocity at position #1. The minimum vibration velocity is obtained at 2.42MS, which is -0.08082131m/S. After that, the vibration velocity continues to fluctuate and the fluctuation period increases significantly.

(3) From the velocity time-history curve in the Y direction at position #3, it can be seen that the vibration changes occur 3MS after the action of blasting. The maximum vibration amplitude is reached at 3.94MS after the detonation of the explosive and the minimum vibration velocity is reached at 4.54MS, with the maximum and minimum vibration velocities being 0.04928285m/S and -0.07120735m/S respectively. Then, the blasting vibration produces multi-stage fluctuations and the amplitude of vibration velocity change in each stage decreases.

(4) At measuring point #4, it remains static in the first 4MS under the blasting action. When the blasting action continues, the maximum vibration amplitude is achieved at 5.76MS after the blasting action with a value of 0.0304699m/S. The minimum vibration velocity is achieved at 6.66MS, which is 0.05145128m/S. After the maximum amplitude appears, it shows a state of continuous fluctuations.

Through the analysis of the velocity time-history curve of measuring

74

points #1~#4 in the Y direction, the peak vibration velocity of the particle at measuring point #1 is still the maximum compared with that at the other measuring points. The amplitude decrease at both measuring point #1 and measuring point #2 in the Y direction exceeds 50% compared with that in the X direction. However, the fluctuation period of blasting vibration velocity in the Y direction increases significantly with the increase of time while the attenuation amplitude decreases gradually.

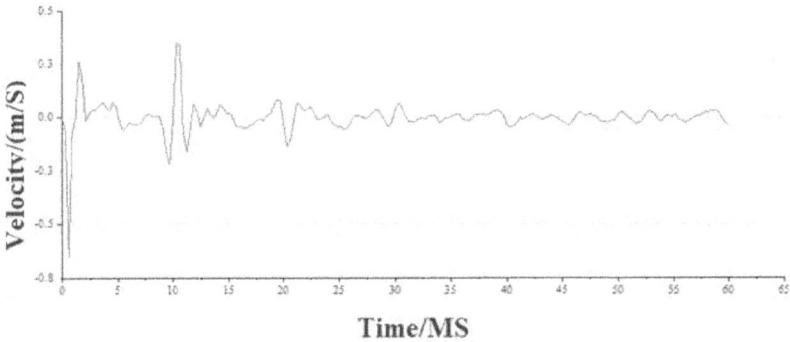

Fig.4.12 Velocity curve in the Z direction at position 1, $\bar{R} = 1$

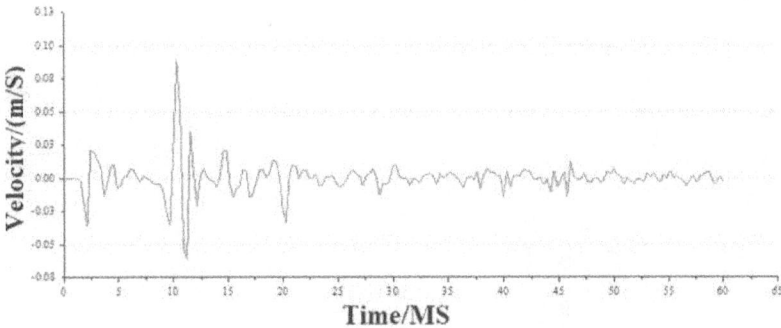

Fig.4.13 Velocity curve in the Z direction at position 2, $\bar{R} = 5$

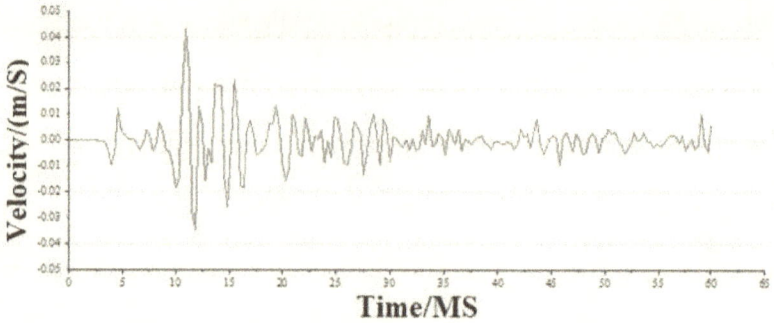

Fig.4.14 Velocity curve in the Z direction at position 3, $\overline{R} = 10$

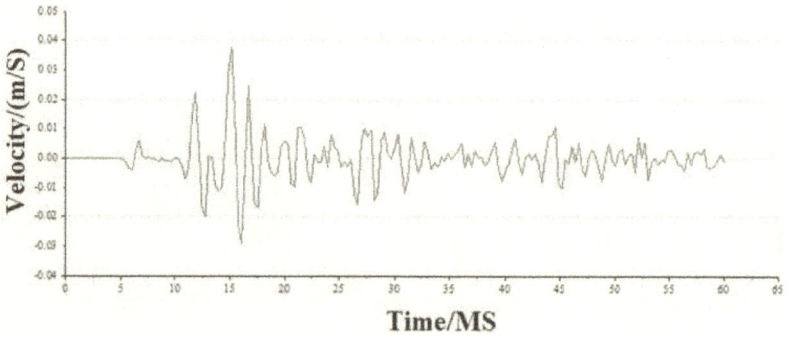

Fig.4.15 Velocity curve in the Z direction at position 4, $\overline{R} = 15$

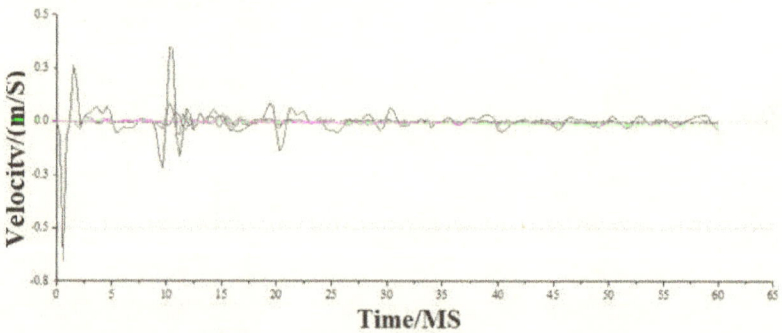

Fig.4.16 Velocity curves in the Z direction at positions #1 to #4

Based on the analysis of Figures 4.12~4.16, the following conclusions can be made.

(1) When the proportional distance is 1, the blasting vibration velocity completes the first fluctuation within 3MS after the detonation of the explosive and the minimum vibration velocity is -0.6518255m/S at 0.61MS. The second vibration fluctuation occurs within 8~12MS and the maximum blasting vibration amplitude is 0.3487898m/S. After that, the vibration velocity decreases, showing a continuous fluctuation state.

(2) At position #2, the blasting vibration velocity fluctuates slightly within 0~5MS and then the amplitude increases. The maximum vibration velocity is achieved at 10.3MS after detonation, with a value of 0.08841193m/S, which is 74.3% higher than the blasting vibration velocity at measuring point #1 in the Y direction, and the minimum vibration velocity is achieved at 11.21MS, with a value of -0.06107447m/S. After 15MS, the blasting vibration amplitude decreases and the vibration velocity tended to stabilize.

(3) At position #3, the blasting vibration velocity remains stable within the first 3MS and then accelerates gradually. The maximum vibration amplitude is reached at 10.9MS, with the maximum value of 0.04329116m/S and the minimum value of -0.03464633m/S. Afterwards, the fluctuation amplitude of vibration velocity decreases, showing a stable fluctuation state after 20MS.

(4) At position #4, the vibration velocity increases gradually from 5MS and the maximum vibration amplitude is 0.0377477 m/S at 15.15MS. The minimum value of vibration velocity is -0.02920212m/S at 16.05MS after

77

detonation. Afterwards, the vibration velocity continues to attenuate and oscillate.

Through the analysis of the velocity time-history curves in the Z direction at measuring points #1~#4, it can be seen that the peak vibration velocity of the particle at the proportional distance of 1 is much larger than that of the other measuring points. This conclusion is consistent with the vibration velocity propagation law in the X and Y directions, but can be clearly seen that the maximum vibration velocity in the Z direction is obtained not at the first fluctuation, but in the time period of 8~18MS after the detonation of the explosive.

4.3 Simulation at Further Points

Based on the model, the calculation results of measuring points #5~#8 are extracted and the velocity time-history curves in the X, Y, Z directions are analyzed to study the propagation law of blasting vibration velocity with the proportional distances of $\bar{R}=20, \bar{R}=30, \bar{R}=40, \bar{R}=50$.

An analysis of the velocity time-history curves in the X direction at measuring points #5~#8 in Figures 4.17~ 4.21 reveals the following.

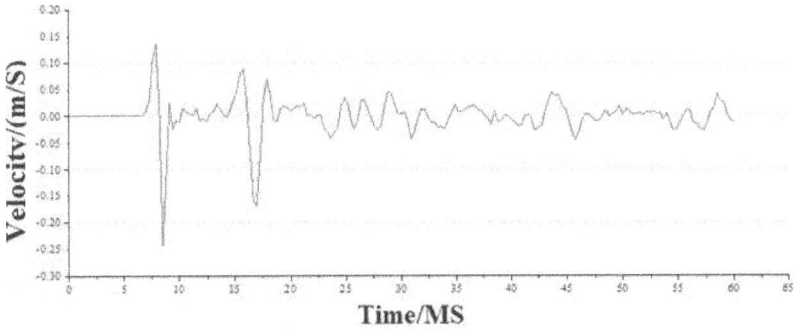

Fig.4.17 Velocity curve in the X direction at position 5, $\bar{R} = 20$

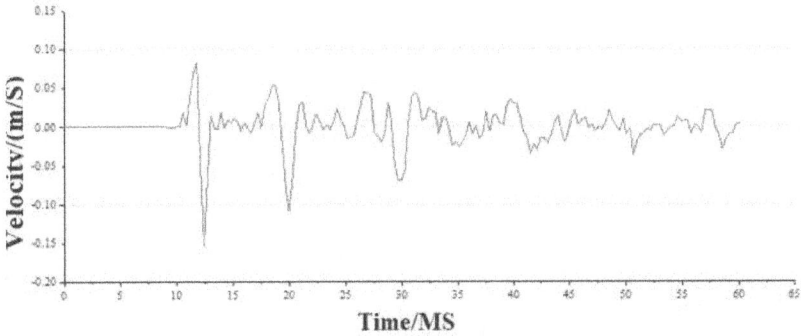

Fig.4.18 Velocity curve in the X direction at position 6, $\bar{R} = 30$

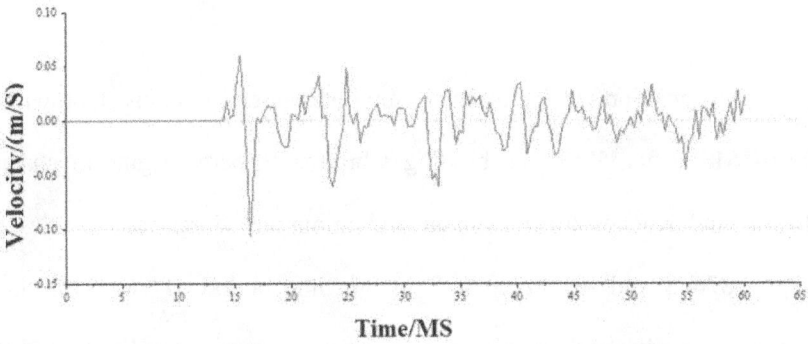

Fig.4.19 Velocity curve in the X direction at position 7, $\bar{R} = 40$

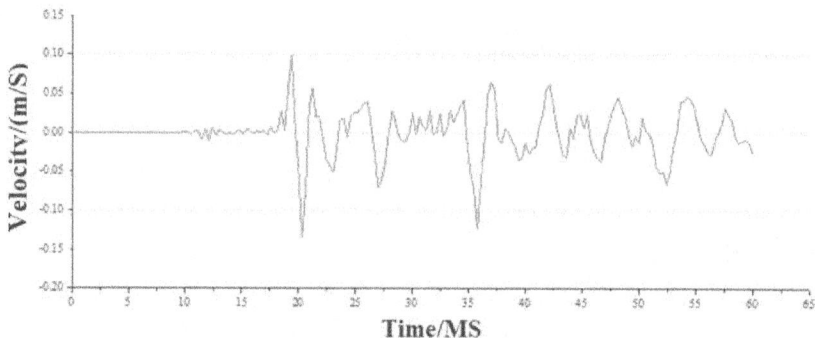

Fig.4.20 Velocity curve in the X direction at position 8, $\overline{R} = 50$

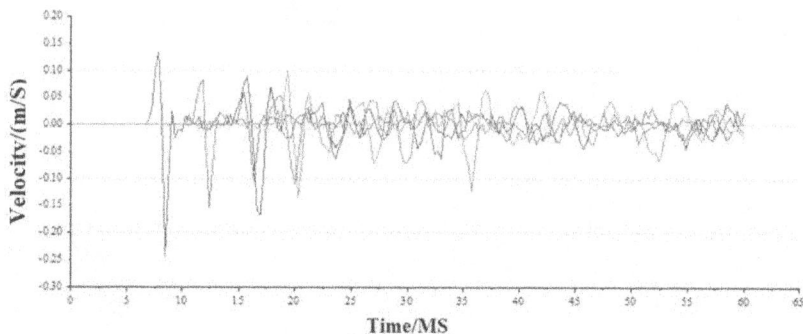

Fig.4.21 Velocity curve in the X direction at position #5 to #8

(1) At position #5, the blasting vibration velocity remains static within the 0-5MS. After 5MS, the blasting vibration velocity begins to change sharply and reaches the maximum and minimum vibration velocities of 0.1343187m/S and -0.2447007m/S in the time period of 5MS~10MS. A significant secondary fluctuation occurs again in the time period of 13~20MS, after which the vibration velocity decreases gradually. The blasting vibration amplitude decreases, showing continuous fluctuations.

80

(2) At position #6, the velocity time-history curve in the X direction remains static within the first 7MS of calculation. subsequently, the maximum vibration velocity is reached at 11.81MS after blasting, with a value of 0.08379595m/S and an attenuation amplitude of 38.5%. The minimum value of vibration velocity is obtained at 12.42MS, which is -1.10115×10^{-10}m/S. It is evident that after reaching the maximum vibration amplitude, the vibration velocity sees an increasing fluctuation period compared with that at measuring point #5. After 30MS, the fluctuation tends to stabilize.

(3) At position #7, the velocity time-history curve in the X direction fluctuates slightly from 10MS. Within 10MS~15MS, the amplitude of blasting vibration velocity gradually increases, reaching the maximum vibration velocity of 0.05979042m/S at 15.45MS. The minimum vibration velocity of -0.1052437m/S is observed at 16.36MS after detonation. The vibration velocity continues to oscillate thereafter, but with a smaller attenuation amplitude.

(4) At position#8, the blasting vibration velocity in the X direction begins to fluctuate slowly around 10MS. The vibration amplitude increases at 16MS after detonation, reaching its maximum in the time period of 17.5~22.5MS, with the maximum and minimum values of 0.09936953m/S and -0.1338769m/S respectively. After reaching the maximum vibration amplitude, the vibration velocity decreases slightly, but the overall amplitude variation is small.

At measuring points #5~#8, the attenuation waveforms of blasting vibration velocity in the X direction are similar, but the variation amplitude between measuring points is small. With the increase of time, the time when the blasting vibration velocity begins to fluctuate continues to delay. With the increase of proportional distances, the number of vibration velocity fluctuations increases while the amplitude variation between successive fluctuations decreases gradually.

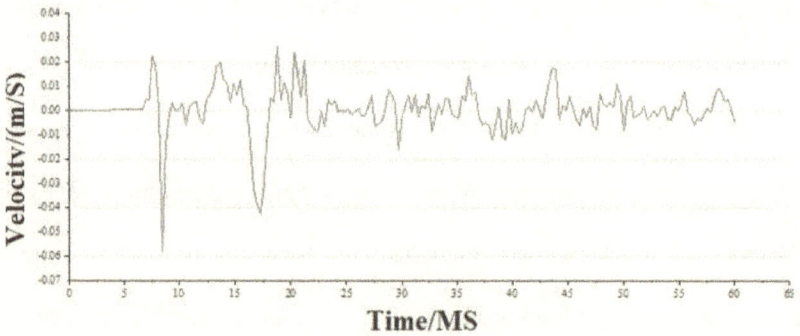

Fig.4.22 Velocity curve in the Y direction at position 5, $\bar{R} = 20$

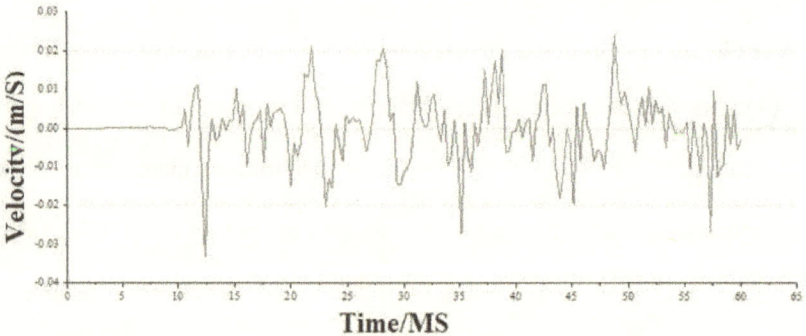

Fig.4.23 Velocity curves in Y direction at position 6, $\bar{R} = 30$

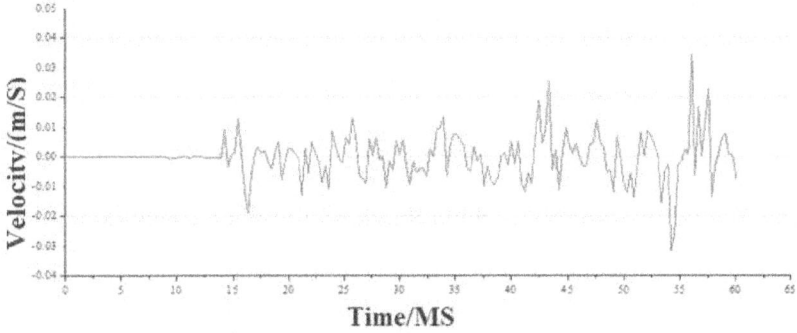

Fig.4.24 Velocity curve in the Y direction at position 7, $\overline{R} = 40$

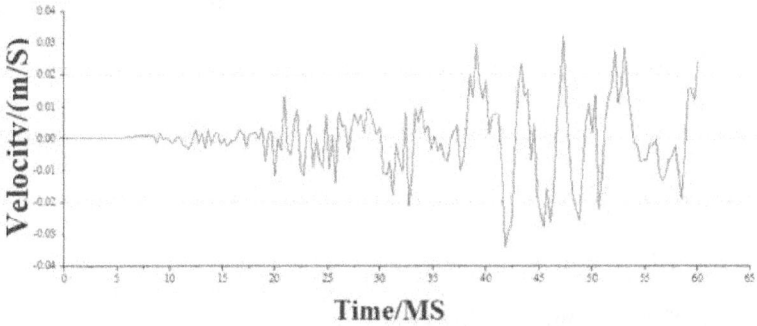

Fig.4.25 Velocity curve in the Y direction at position 8, $\overline{R} = 50$

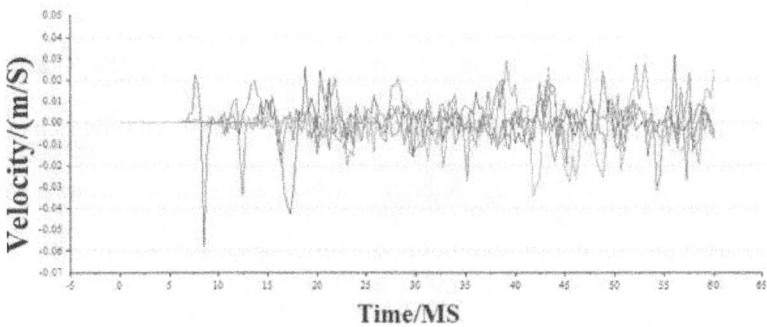

Fig.4.26 Velocity curves in the Y direction at position 5 to 8

An analysis of the blasting vibration velocity in the Y direction at measuring points #5~#8 in Figures 4.22~ 4.26 yields the following results.

(1) At measuring point #5, the blasting vibration velocity remains static within 0~5MS, and small fluctuations begin to occur after 5MS of blasting. Large vibration amplitude changes occur in the period of 5MS~10MS. At 8.48MS after detonation of the explosive, the minimum vibration velocity is reached, with a value of -0.0583329m/S, and the maximum vibration velocity is reached, with a value of 0.02657794m/S at 18.78MS. It is evident that large fluctuations occur again in the period of 10MS~20MS.

(2) At measuring point #6, the blasting vibration velocity begins to change at 8MS, but its change range was very small. At 10MS after the detonation, the blasting vibration amplitude in the Y direction increases gradually. The maximum vibration velocity is reached at 12.42MS, which is 0.02435792m/S, while at 48.78MS, the minimum vibration velocity is reached, which is -0.03322617m/S. The vibration velocity fluctuates uniformly as a whole.

(3) At position #7, the blasting vibration velocity of in the Y direction fluctuates slightly after 10MS and shows a stable fluctuation state in the time period of 14~40MS. A large vibration velocity fluctuation occurs within 40~45MS and the blasting vibration amplitude increased within 55~60MS, reaching the maximum vibration velocity of 0.03493376m/S at 50.05MS and the minimum vibration velocity of -0.03159798m/S at 54.24MS.

(4) At position #8, similar to the situation of position #7, the blasting vibration velocity fluctuates steadily within the first 37MS after the detonation

of the explosive. The vibration amplitude is small, but it increases after 37MS. The maximum vibration velocity is reached at 51.2MS, with a value of 0.03239744m/S and the minimum vibration velocity is reached at 49.99MS, with a value of -0.03429489m/S.

The variation law of blasting vibration velocity in the Y direction of measuring points #5~#8 is significantly different from that in the X direction of the same measuring points. With the increase of proportional distance, the vibration velocity variation amplitude presents a stable fluctuation state after the detonation of the explosive, while the vibration amplitude increases after 40MS. The greater the proportional distance, the longer the time it takes to obtain the particle peak vibration velocity.

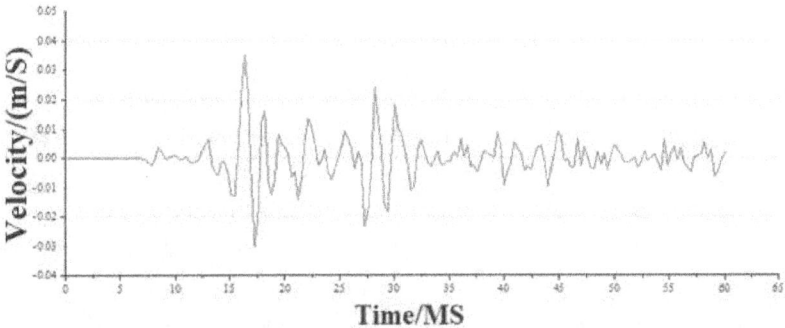

Fig.4.27 Velocity curve in the Z direction at position 5, $\overline{R} = 20$

Based on Figures 4.27-4.31, an analysis is conducted on the velocity time-history curves at measuring points #5~#8 position yields the following results.

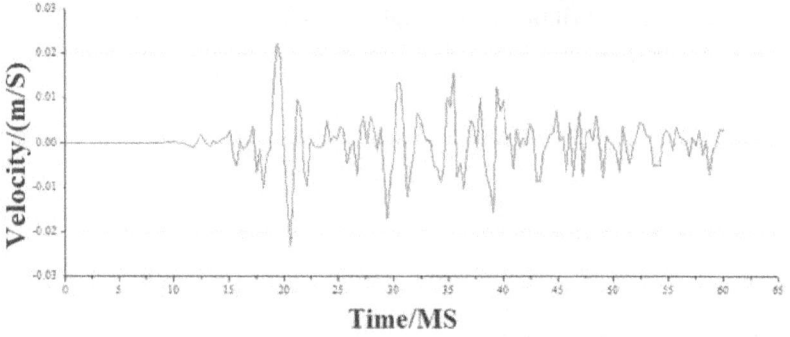

Fig.4.28 Velocity curve in the Z direction at position 6, $\overline{R} = 30$

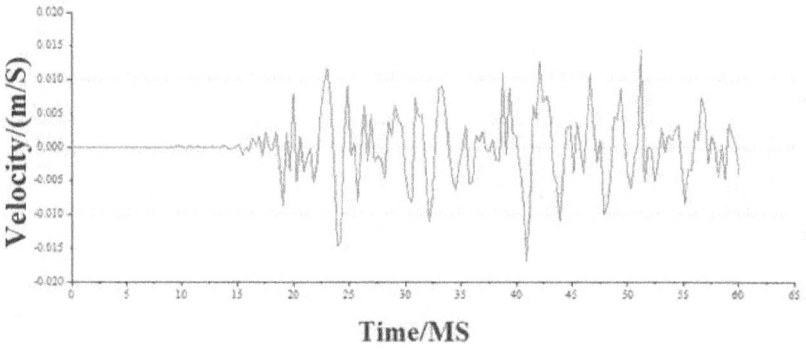

Fig.4.29 Velocity curve in the Z direction at position 7, $\overline{R} = 40$

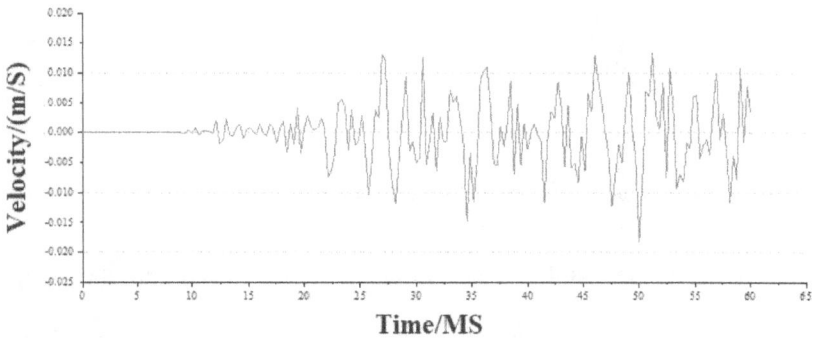

Fig.4.30 Velocity curve in the Z direction at position 8, $\overline{R} = 50$

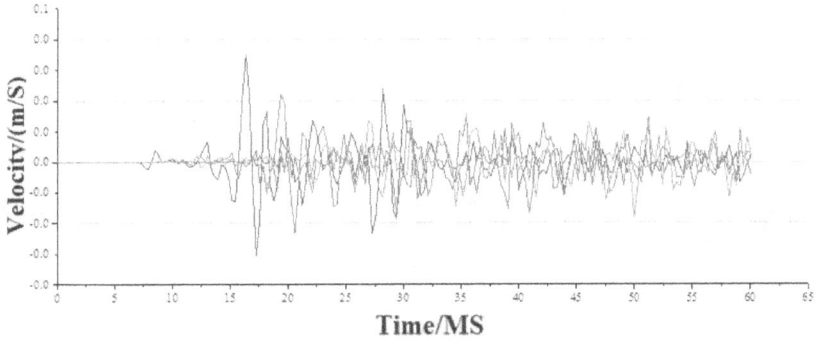

Fig.4.31 Velocity time-history curves in the Z direction at positions #5 to #8

(1) At measuring point #5 and when the proportional distance $\bar{R}$=20, the blasting vibration velocity remains static within the first 5MS after detonation and the vibration amplitude is small within 5MS~13MS. After 15MS, the blasting vibration velocity changes rapidly. The maximum vibration velocity is reached at 16.36MS after detonation, with a value of 0.03520335m/S, and the minimum blasting vibration is reached at 17.27MS, with a value of -0.03030432m/S. After 20MS, the amplitude decreases and the vibration velocity fluctuates again within 25MS~30MS and then continues to fluctuate after 35MS.

(2) At measuring point #6, the blasting vibration velocity remains unchanged within 0~8MS and then there was a slight fluctuation. The vibration velocity increases at 15MS, reaching its maximum value of 0.02226768m/S at 19.39MS after detonation. Compared with measuring point #5, the amplitude decreases by 37.1%. The minimum vibration velocity is

reached at 20.6MS, with a value of -0.02319745m/S. Afterwards, the amplitude decreases and the vibration velocity continues to fluctuate.

(3) At measuring point #7, the variation in vibration amplitude is relatively small within the first 10MS after blasting. After 15MS, the amplitude of blasting vibration velocity increases gradually. After 22MS, the vibration velocity fluctuations become relatively uniform in magnitude. The minimum vibration velocity occurs at 40.9MS, with a value of -0.0168605m/S, and the maximum vibration velocity occurs at 51.2MS, with a value of 0.01432059m/S.

(4) At measuring point #8 and when the proportional distance $\bar{R}$=50, the blasting vibration velocity remains static within 8MS after detonation. During the period of 10MS-20MS, the amplitude of blasting vibration velocity shows minor variations. After 20MS, velocity starts to change greatly, with relatively uniform vibration amplitude. The maximum and minimum values of the vibration velocity occur in 45MS~55MS, with the maximum and minimum values of particle vibration velocity being 0.01335062m/S and -0.01825473m/S respectively.

By examining the velocity time-history curved of measuring points #5~#8 in the Z direction, it is observed that the vibration velocity attenuation trend is similar to that in the Y direction at the same locations. When the proportional distance falls within the range of 20~30, the vibration velocity maintains a waveform similar to that observed at measuring points #1~#4. However, when the proportional distance is in the range of 40~50, no obvious vibration in vibration amplitude is observed. The vibration velocity fluctuates only slightly within the first 20MS after detonation. After 20MS, the amplitude increases and the velocity changes uniformly, with the extreme values occurring around 60MS.

Table 4.1 The peak time of blasting vibration velocity in the X direction at each measuring point

Measuring point	#1	#2	#3	#4	#5	#6	#7	#8
Time to reach maximum vibration (MS)	0.61	2.12	3.94	6.06	7.88	11.81	15.45	19.39
Time to reach minimum vibration (MS)	0.91	2.42	4.54	6.66	8.48	12.42	16.36	20.29

Table.4.2 The peak time of blasting vibration velocity in the Y direction at each measuring point

Measuring point	#1	#2	#3	#4	#5	#6	#7	#8
Time to reach maximum vibration (MS)	0.61	2.12	3.94	5.76	18.78	12.42	56.05	41.80
Time to reach minimum vibration (MS)	1.51	2.42	4.54	6.66	8.48	48.78	54.24	47.26

Table.4.3 The peak time of blasting vibration velocity in the X direction

at each measuring point

Measuring point	#1	#2	#3	#4	#5	#6	#7	#8
Time to reach maximum vibration (MS)	10.30	10.30	10.90	15.15	16.36	19.39	51.20	51.20
Time to reach minimum vibration (MS)	0.61	11.21	11.81	16.05	17.27	20.60	40.90	49.99

The time at which the blasting vibration velocities reach the peak values in all directions at measuring points #1~#8 is summarized and analyzed in Tables 4.1~4.3.

By analyzing the blasting vibration velocity time-history curves at positions #1~#8 in the X, Y and Z directions and according to Tables 4.1~4.3, the following results can be obtained.

When the proportional distance is 1, the blasting vibration velocities in the X and Y directions reach the maximum vibration amplitude within 1ms, and the minimum vibration velocity in the Z direction is reached within 1MS, indicating that the blasting vibration velocity in each direction has a sharp change at measuring point #1. When the proportional distance is 5~10, the blasting vibration velocity in the X and Y directions reaches the maximum within 5MS, while the peak vibration velocity in the Z direction is obtained after 10MS. When the proportional distance is within the range of 15~30, the maximum blasting vibration amplitudes in all directions are achieved within 20MS. However, a comparison reveals that the time taken to reach the maximum vibration velocity at measuring point #5 in the Y direction is later than that at measuring point #6 in the Y direction, indicating that when the measuring point is within the proportional distance, the time taken to reach the particle peak vibration velocity in the Y direction does not monotonically and continuously increase with the increase of the proportional distance. When the proportional distance is greater than 30, the blasting vibration velocity peak in the X direction still shows continuous attenuation, but the blasting vibration velocities in the Y direction and Z direction fluctuate slightly around 20MS after detonation, with their peak values occurring after 50MS. Therefore, when the blasting vibration waves propagate to this distance range, the variations of blasting vibration velocity in the Y and Z directions lack a distinct attenuation process, and the peak particle vibration velocity occurs later than that in the X direction at the same measuring point.

4.4 Analysis of Maximum and Minimum Velocities

Based on the above analysis, this section further explores the relationship between the maximum and minimum blasting vibration velocities in the X, Y and Z directions and the proportional distance at each measuring point. By using Origin graphing and analysis software, the attenuation patterns are visually represented as shown in Figure 4.32.

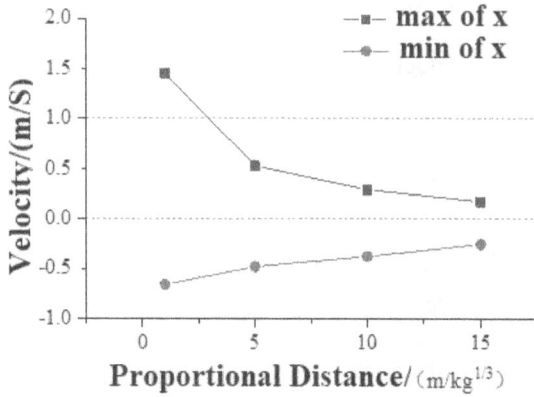

Fig.4.32 Variations of maximum and minimum values of blasting vibration in the X direction with proportional distance at measuring points # 1 to # 4

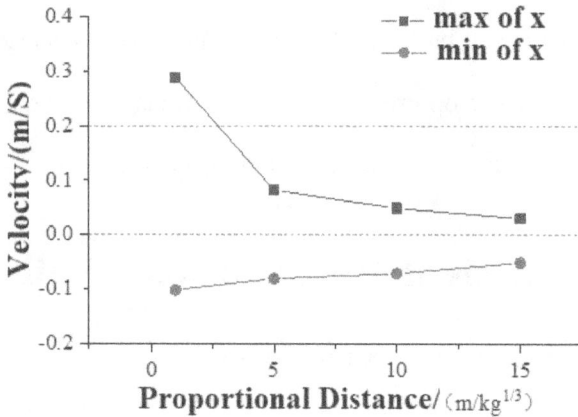

Fig.4.33 Variations of maximum and minimum values of blasting vibration in the Y direction with proportional distance at measuring points # 1 to # 4

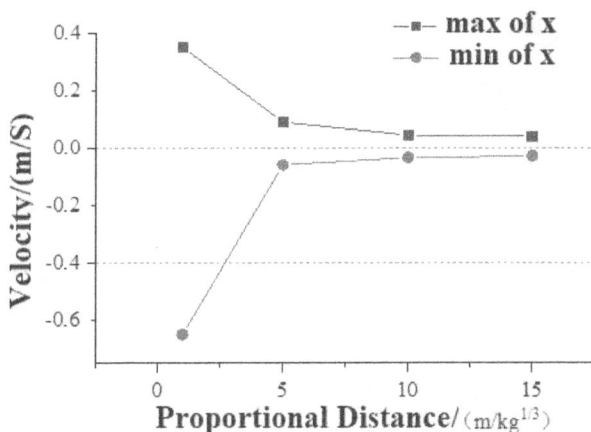

Fig.4.34 Variations of maximum and minimum values of blasting vibration in the Z direction with proportional distance at measuring points # 1 to # 4

From Figures 4.32~4.34, it can be seen that the maximum and minimum values of blasting vibration velocity in all directions are declining with the increase of proportional distance. When the proportional distance is 1, the absolute difference between the maximum and minimum values of blasting vibration velocity in the X direction is the largest, reaching 0.796659m/S, followed by that in the Z direction, with the maximum difference being 0.3030362m/S. When the proportional distance is 5, the maximum values of velocity in the X, Y and Z directions show a rapid downward trend, and the variation ranges of velocity are 63.7%, 72.4% and 74.3% respectively. The decrease ranges of the minimum values of blasting vibration (taking the absolute value) are 37.3%, 20% and 90.8%. It is evident that the attenuation ranges of the maximum and minimum values of blasting vibration in the Z direction are greater than those in the X and Y directions, and that the variation range of the minimum value of vibration velocity in the Z direction is extremely rapid. When the proportional distance is greater than 5, the variations of the maximum and minimum blasting vibration velocities tend to stabilize and the maximum and minimum blasting vibration velocities in all directions show similar trends.

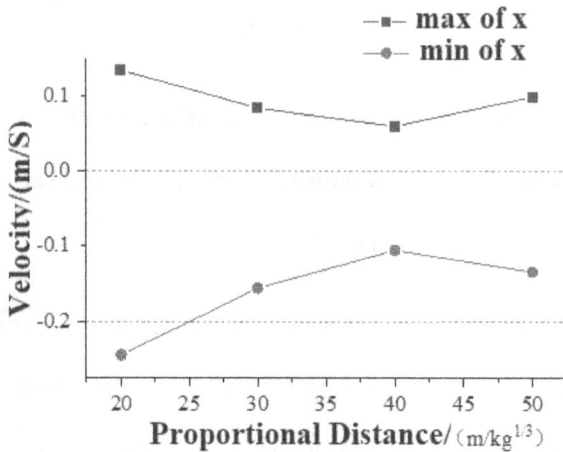

Fig.4.35 Variations of maximum and minimum values of blasting vibration in the X direction with proportional distance at measuring points # 5 to # 8

As shown in Figures 4.35~4.37, as the proportional distance at measuring points #5~#8 increases, the maximum and minimum values of blasting vibration velocity in each direction do not exhibit significant amplitude variations. The absolute values of the maximum and minimum velocities at different proportional distances show only slight differences and do not decay monotonously with the increase of proportional distance.

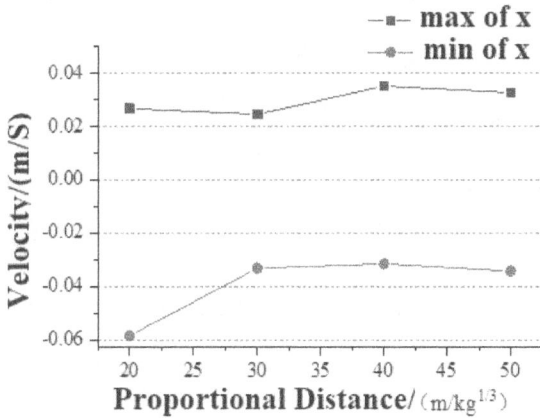

Fig.4.36 Variations of maximum and minimum values of blasting vibration in the Y direction with proportional distance at measuring points # 5 to # 8

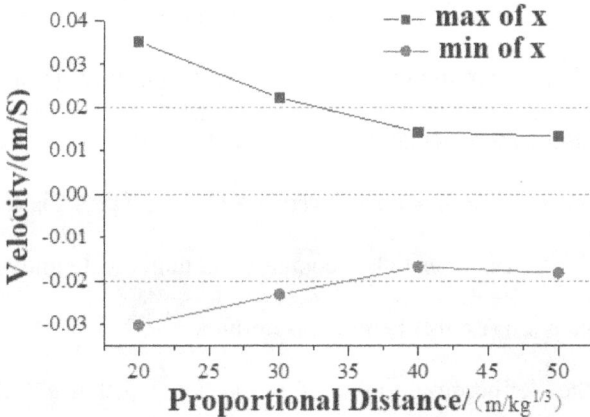

Fig.4.37 Variations of maximum and minimum values of blasting vibration in the Z direction with proportional distance at measuring points # 5 to # 8

When the proportional distance is greater than 30, the blasting vibration amplitudes in all directions demonstrate certain fluctuations. When the proportional distance is 40, the maximum value of particle vibration velocity in the Y direction shows a slight upward fluctuation before attenuating slowly again. When the proportional distance is 50, the absolute values of the maximum and minimum values of blasting vibration velocity in the X direction exhibit an increasing trend and the absolute values of the minimum values of blasting vibration velocity in the Y and Z directions also show a slight increase.

In engineering blasting, the blasting vibration amplitude generally refers to the maximum vibration velocity of the vibration wave. As one of the critical indicators for assessing blast-induced vibration intensity, it plays an important role in studying the propagation law of blasting vibration. China's current blasting regulations commonly adopt the empirical blasting vibration formula, Sadovsky formula, to calculate the peak value of particle vibration velocity. When the numerical simulation software ANSYS/LS-DYNA is used for rock blasting analysis, it is essential to compare the numerical simulation results with the calculation results of empirical formula.

From the definition formula of proportional distance (3.5) and the expression of Sadovsky formula (4.1), the relationship between the peak vibration velocity of particles and proportional distance can be obtained as follows:

$$V = K(\bar{R})^{-\alpha} \tag{4.2}$$

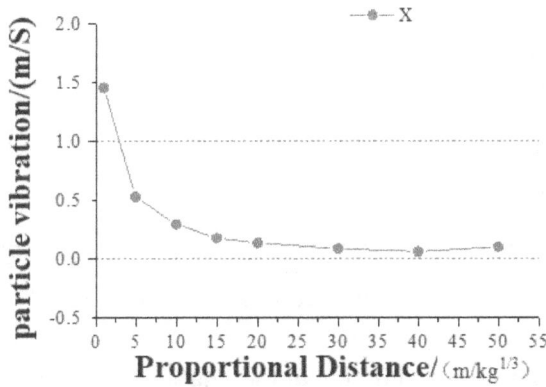

Fig.4.38 Numerical simulation of the relationship between the peak value of particle velocity in the X direction and the proportional distance

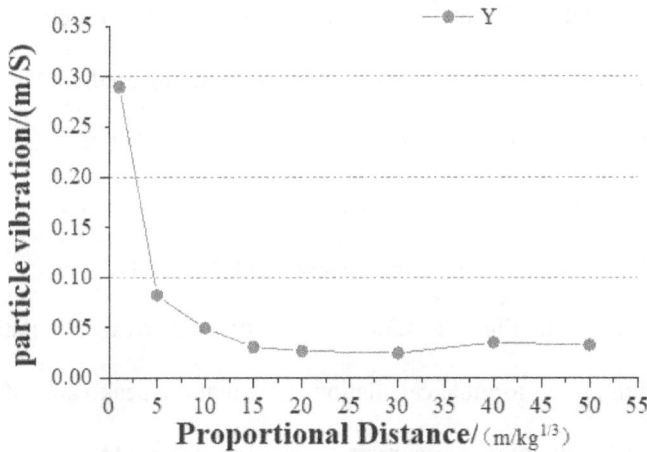

Fig.4.39 Numerical simulation of the relationship between the peak value of particle velocity in the Y direction and the proportional distance

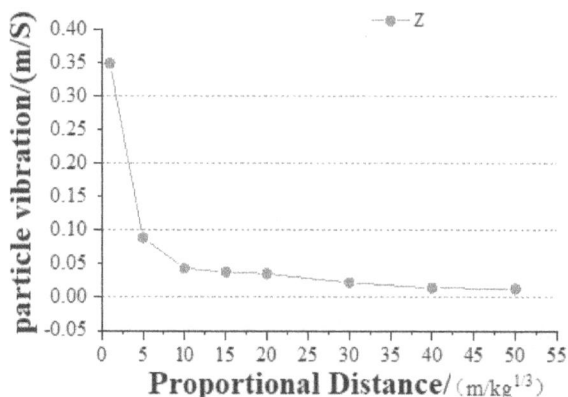

Fig.4.40 Numerical simulation of the relationship between the peak value of particle velocity in the Z direction and the proportional distance

Where, V is the particle peak vibration velocity; $\bar{R}$ is the proportional distance; K and α are the site parameters and attenuation coefficient related to geological conditions, respectively. In this study, taking medium hard rock as an example, taking K=200, α=1.6, the relationship between the blasting vibration velocity peak and the proportional distance can be obtained by using MATLAB data analysis software, as shown in Figure 4.41.

According to Figures 4.38~4.41, both numerical simulation and Sadowski empirical formula calculation show that the peak value of blasting vibration velocity follows a similar trend with respect to the proportional distance. This consistency simultaneously confirms that it is feasible to solve the peak value of blasting vibration velocity by inputting engineering data from the blasting site into the numerical simulation.

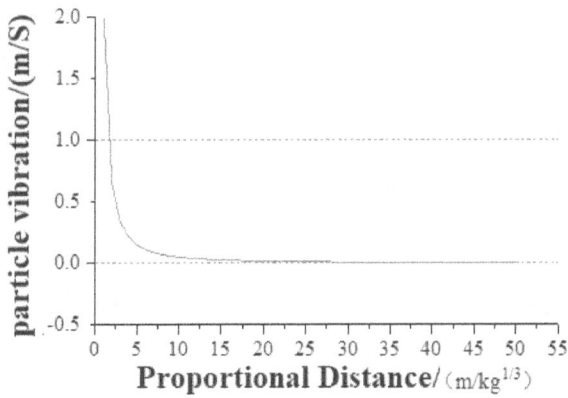

Fig.4.41 Relationship between peak particle velocity and proportional distance solved by Sadowski formula

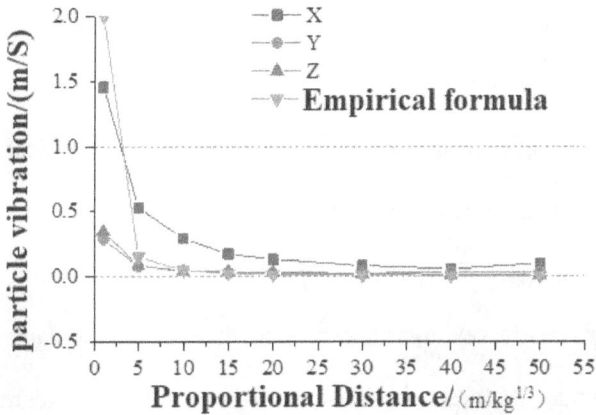

Fig.4.42 Comparison between numerical simulation results and theoretical calculation results

Through comparative analysis, it can be seen that when the proportional distance is 1~5, the blasting vibration amplitude decreases sharply in a short time and the vibration velocity attenuates rapidly. This area can be classified as the near blasting area, namely AB section. When the proportional distance is 5~10, the blasting vibration peak value decreases slightly and the velocity attenuation slows down gradually. This area can be classified as the middle blasting area, namely BC section. When the proportional distance is greater than 10, the blasting vibration amplitude changes very little with only slight fluctuations and the vibration velocity attenuation tends to stabilize. This area can be classified as the far blasting area, namely CD section.

In addition, in the far region of CD section where the proportional distance is greater than 10, the curve representing the relationship between the peak particle velocity calculated by Sadovsky formula and the proportional distance is almost parallel to the distance axis. However, the curve derived from numerical simulation results shows a gradual attenuation process. Therefore, in engineering practice, relying solely on empirical formula to calculate the peak value of blasting vibration velocity in the far blasting area carries certain risks. To accurately predict the peak vibration velocity of blasting particles in key areas and reasonably evaluate the damage effect of structures under the action of blasting, it is essential to integrate the field monitoring data, experimental analysis and numerical simulation.

4.5 Study on Blasting Vibration Frequency

In recent years, studies by Chinese scholars and field engineers on the attenuation law of blasting vibration has revealed that blasting vibration frequency is also an important factor affecting the safety of blasting vibration. The frequency of blasting seismic waves refers to the physical quantity characterizing the rapidity of harmonic motion of rock mass particles under the action of blasting, which is related to the design parameters of the explosive, the distance between the blasting center and the physical and mechanical properties of the rock mass.

The blasting vibration frequency of the seismic waves presents a multi-band waveform with a wide distribution range. When the waves propagate in the rock mass medium, the amplitude and frequency of the blasting vibration continue to decrease with the increase of the proportional distance, while the duration of the blasting vibration lengthens. The blasting seismic waves of the high frequency components attenuate rapidly while the blasting seismic waves of the low frequency components attenuate slowly. The low-frequency blasting seismic waves become the main component at locations farther from the blast source.

The different vibration frequencies generated by blasting cause different degrees of damage to structures. According to the resonance principle, when the natural vibration frequency of a structure is close to the blasting vibration frequency, resonance occurs between the structure and the seismic waves, amplifying the vibration energy and aggravating the structural damage. Therefore, it is necessary to analyze and study the blasting vibration frequencies at different distances from the blasting source when discussing the propagation laws of blasting vibration.

Based on numerical simulations, this study takes measuring point #1 as the characteristic point in the near area of blasting vibration (Fig.4.1), measuring points #2 and #3 as the characteristic points in the middle area of blasting vibration, measuring points #4, #5, #6, #7 and #8 as the characteristic points in the far area of blasting vibration. By using the spectrum analysis function in MATLAB data analysis software, the blasting vibration velocity data undergo Fourier transformation to convert time-domain signals into frequency-domain signals so as to further analyze the attenuation law of blasting vibration frequency under different proportional distances. The specific results are as follows:

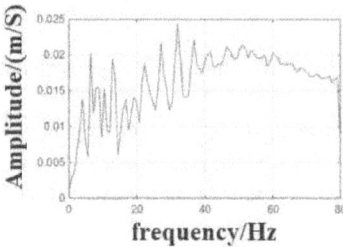

(a) #1 X direction FFT

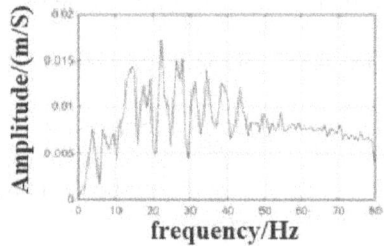

(b) #2 X direction FFT

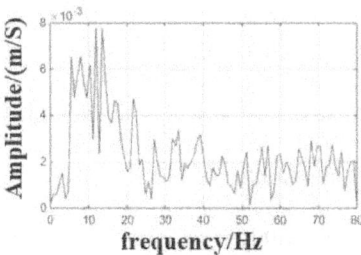

(c) #3 X direction FFT

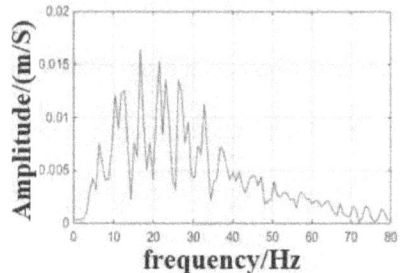

(d) #4 X direction FFT

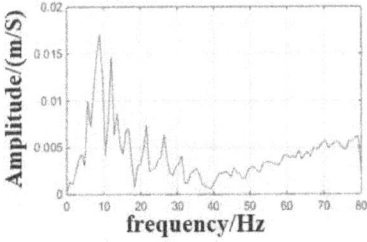

(e) #5 X direction FFT

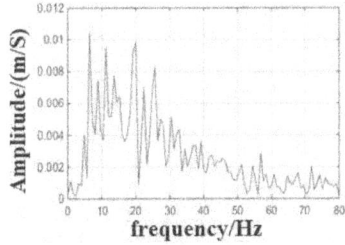

(f) #6 X direction FFT

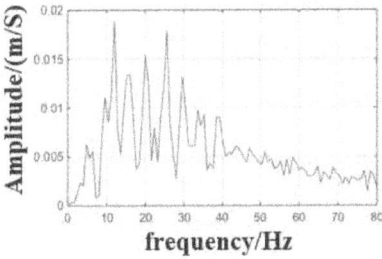

(g) #7 X direction FFT

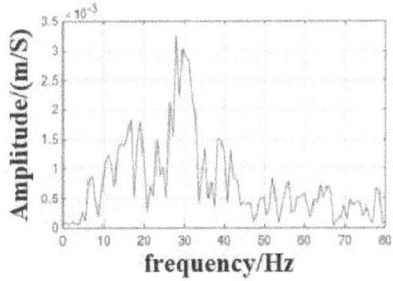

(h) #8 X direction FFT

Fig.4.43 X direction FFT of blasting vibration

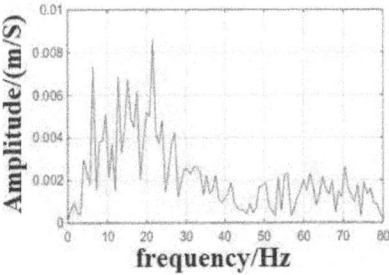

(a) #1 Y direction FFT

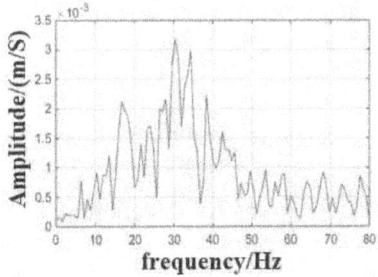

(b) #2 Y direction FFT

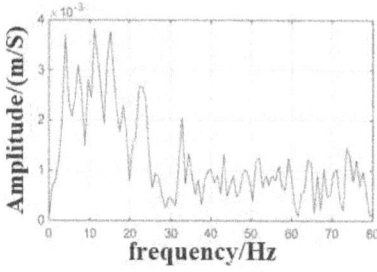

(c) #3 Y direction FFT

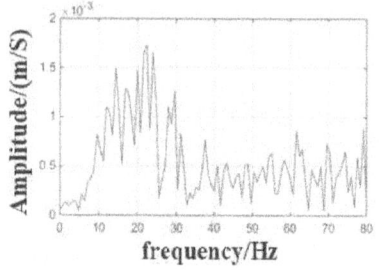

(d) #4 Y direction FFT

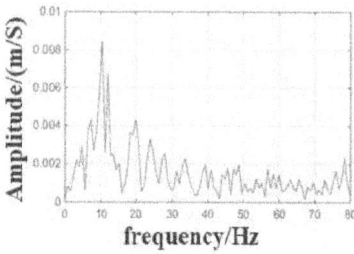

(e) #5 Y direction FFT

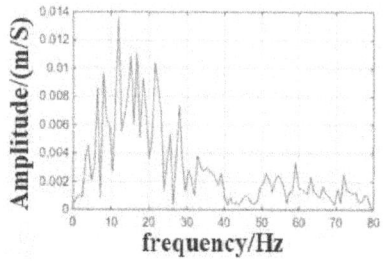

(f) #6 Y direction FFT

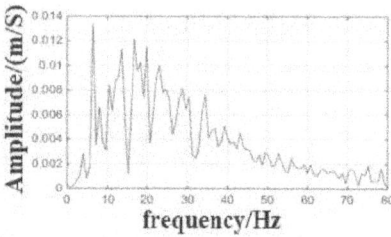

(g) n#7 Y direction FFT

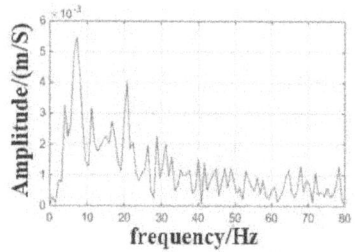

(h) #8 Y direction FFT

Fig.4.44 Y direction FFT of blasting vibration

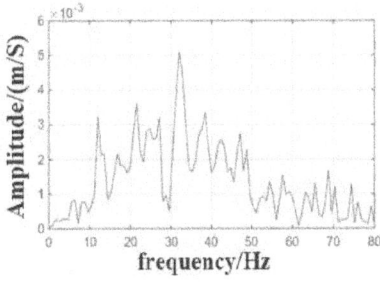

(a) #1 Z direction FFT

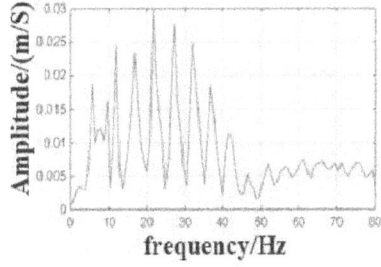

(b) #2 Z direction FFT

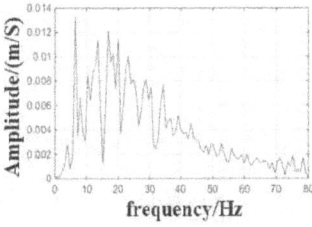

(c) #3 Z direction FFT

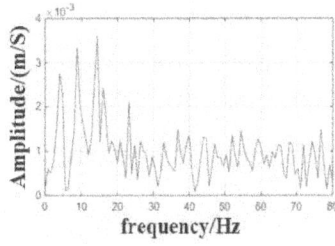

(d) #4 Z direction FFT

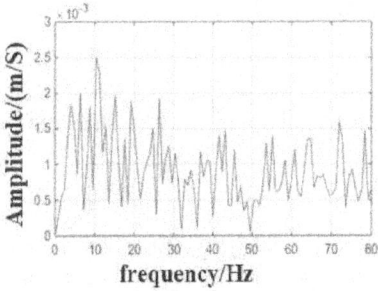

(e) #5 Z direction FFT

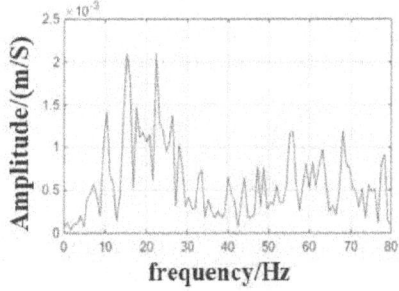

(f) #6 Z direction FFT

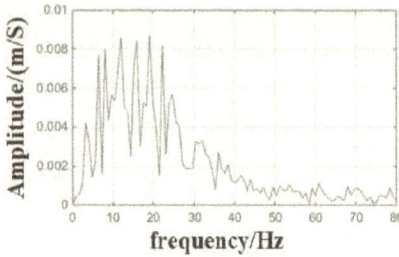

(g) #7 Z direction FFT   (h) #8 Z direction FFT

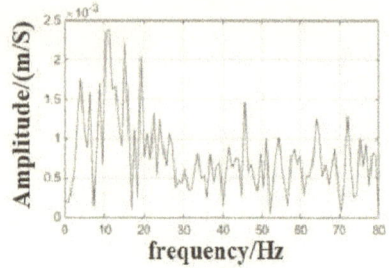

Fig.4.45 Z direction FFT of blasting vibration

As can be seen from Figures 4.43~4.45, the FFT amplitude transformation curves in the X, Y and Z directions indicate that as the proportional distance increases, the distribution of blasting vibration spectrum narrows gradually.

When the proportional distance is 1 (i.e. the near blasting vibration area), the blasting vibration spectrum in the X direction exhibits the widest distribution range, characterized by multiple peaks and broad distribution of blasting vibration frequency. The blasting vibration spectrum in the Y direction is distributed in the range of 15 to 40Hz, while the blasting vibration spectrum in the Z direction is mostly distributed in the range of 10 to 50Hz, with a small part extending to the range of 50 to 70Hz. When the proportional distance is 5~10 (i.e. the middle blasting vibration are), the blasting vibration spectrum gradually transits from multiple dominant frequencies to a single dominant frequency, and the distribution range of blasting vibration spectrum narrows. The high-frequency components decrease gradually whereas the low-frequency components increase. When the proportional distance is greater than 10 (i.e. the far blasting vibration area), the blasting seismic wave frequency exhibits a stable single dominant frequency state, with a narrow frequency distribution range predominantly composed of low-frequency components.

According to the above numerical simulation results, the numerical simulation of blasting vibration propagation velocity and blasting vibration frequency at each measuring point is shown in Table 4.4.

Table 4.4 Numerical simulation data of blasting vibration in each area

Pts	M (Kg)	Dis (m)	Pro. Dis.	zone	Vibration (m/S) X -Direc.	Y -Direc.	Z -Direc.	Freq. (Hz) X -Direc.	Y -Direc.	Z -Direc.
#1	2.5	1.36	1	close	1.456	0.289	0.349	32	30.4	32
#2	2.5	6.78	5	mid.	0.530	0.082	0.088	22.4	21.6	21.6
#3	2.5	13.57	10	mid.	0.294	0.049	0.043	12	11.52	6.4
#4	2.5	20.35	15	far	0.174	0.030	0.038	28	22.4	14.4
#5	2.5	27.14	20	far	0.134	0.027	0.035	13.6	11.2	10.4
#6	2.5	40.72	30	far	0.084	0.024	0.022	16.8	12	22.4
#7	2.5	54.28	40	far	0.060	0.035	0.014	8.8	6.4	12
#8	2.5	67.86	50	far	0.099	0.032	0.013	6.4	7.2	10.4

The main frequency of blasting vibration refers to the frequency corresponding to the maximum vibration amplitude in the FFT amplitude transformation curve. According to the blasting vibration spectrum in the X, Y and Z directions of the above measuring points, the image data is derived by MATLAB to obtain the main vibration frequency of each point. The software Origin is adopted to describe the relationship between the main vibration frequency of blasting seismic waves and the attenuation of blasting vibration proportional distance, as shown in the following figures:

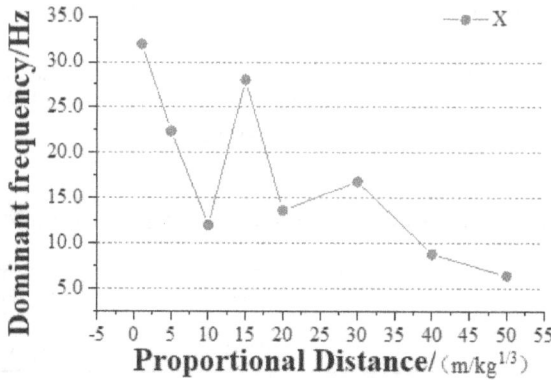

Fig.4.46 Variation of blasting vibration frequency in the X direction

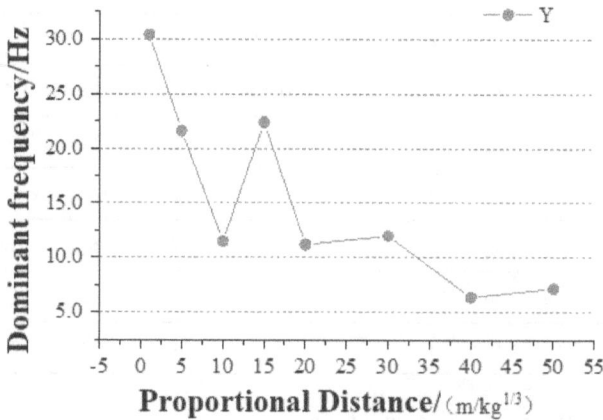

Fig.4.47 Variation of blasting vibration frequency in the Y direction

According to Figures 4.46~4.49, when it is located in the near blasting vibration area (the proportional distance is 1), the main frequencies of blasting vibration in all directions are higher, which are above 30Hz. When the blasting vibration waves propagate to the middle blasting vibration area, the main vibration frequency attenuates rapidly and the low-frequency part of the blasting vibration frequency becomes the main part because of the rapid attenuation of high-frequency and high-velocity rock mass medium in the process of seismic wave propagation.

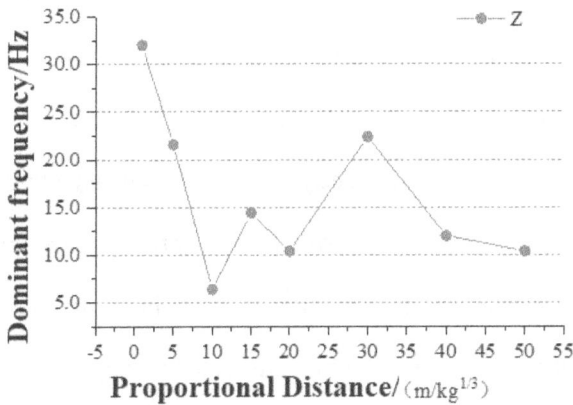

Fig.4.48 Variation of blasting vibration frequency in the Z direction

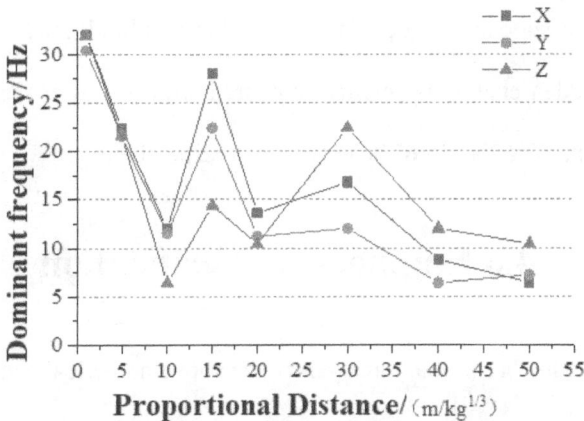

Fig.4.49 Variations of blasting vibration frequency in 3 directions

However, when the proportional distance is within the range of 10~20, the main frequencies of blasting vibration in the X, Y and Z directions produce a sudden change. After the sudden change, the main frequency of blasting vibration in the X direction is the largest, reaching 28Hz. When the proportional distance is within the range of 20~40, the main frequency of blasting vibration fluctuates again and the main frequency of blasting vibration in the Z direction exhibits the largest increase, reaching 22.4Hz. When the blasting vibration waves propagate to the far blasting vibration area (where the proportional distance is greater than 40), the amplitude and main frequency of blasting vibration decrease gradually and tend to stabilize.

At the same time, compared with the two abrupt fluctuations of the main frequency in the far blasting vibration area, the main frequency of blasting vibration in the X, Y and Z directions attenuated rapidly after the first sudden change, with a decrease of 51.43%, 50% and 27.78% respectively. The second abrupt change of main frequency shows a decreasing amplitude compared with the first sudden change. The attenuation amplitude after reaching the value of sudden change is smaller compared with that after the first abrupt change, being 47.62%, 46.67% and 46.43% respectively.

4.6 Engineering Verification

According to the preliminary analysis of the above numerical simulations, the authenticity and accuracy of the results are tested in combination with engineering examples. The numerical simulation results are compared with the measured results in the actual engineering to explore the particle propagation law under the action of blasting. Taking an earth rock blasting project in Yantai as an example, this study conducts an on-site monitoring of the blasting vibration. The monitored data are analyzed and compared with the results of numerical simulations.

The project is located in Yantai City, Shandong Province. The landform is simple. The rock types are mainly limestone and dolomite. The rock hardness coefficient is f=5~10, belonging to rocks with medium hardness or above. This blasting project is to blast and level the backfilling according to the construction plan. The site picture is shown in Figure 4.50.

Fig.4.50 Photo of site construction

The region where the project is located has rapid economic development, developed industry and convenient transportation. The earth rock blasting project will accelerate the regional development and promote the improvement of its economy. According to the geological conditions and practical experience of the blasting project, the blasting depth of the project is determined to be 5~15m, and the blasting quantity is about 150,000 square meters.

The blasting vibration monitoring equipment is used to monitor the vibration characteristics of Yantai blasting earthwork project, and the blasting vibration monitoring data of each measuring point is recorded by the blasting vibration recorder. TC-4850 blasting vibration meter is used to monitor the blasting engineering site.

1. Selection of monitoring points

According to the actual engineering situation, six monitoring points are selected to analyze the blasting vibration velocity and the corresponding proportional distance is calculated according to formula (3.5), as shown in Table 4.4.

Table.4.4 Parameters of selected blasting vibration monitoring points

monitoring point	1	2	3	4	5	6
Distance from blasting center (m)	101	71	112	81	132	191
Amount of explosive (kg)	500	50	100	30	90	90

Proportional distance $(m/kg^{\frac{1}{3}})$	12.73	19.27	24.13	26.07	29.46	42.62

2. Vibration monitoring results

The millisecond blasting technology is adopted in this earthwork blasting project. From the change curve of the blasting vibration velocity recorded by the vibrometer, it can be seen that there is a certain blasting interruption time between each section of blasting vibration velocity waveform, and the effective monitoring values of each monitoring point can be obtained from the on-site engineering situations.

The monitoring data of measuring point 1 are shown in the table below:

Table.4.5 Data of measuring point 1

Chanel	Vib. Max (cm/S)	Time at max (S)	Dominant Freq. (Hz)	Range (cm/S)	Sensitivity (V/m/S)
X	2.19	8.7375	21.74	37.99	26.32
Y	1.10	9.1511	17.39	40.52	24.68
Z	2.41	9.2230	15.87	37.43	26.72

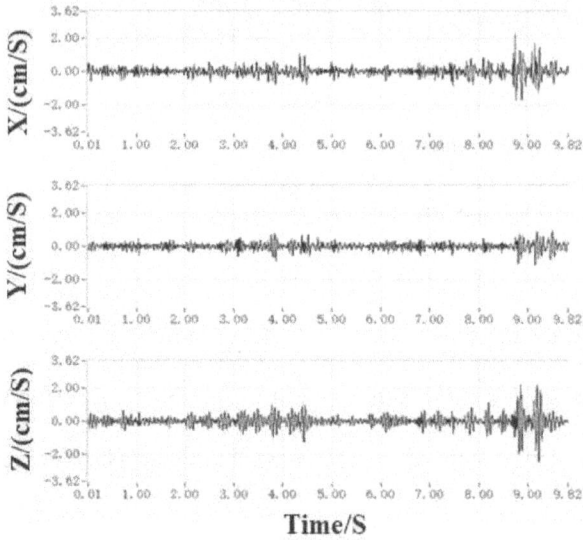

Fig.4.51 Blasting velocities in the X, Y and Z directions at point 1

The monitoring data of measuring point 2 are as follows:

Table.4.6 Data of measuring point 2

Chanel	Vib. Max (cm/S)	Time at max (S)	Main Freq. (Hz)	Range (cm/S)	Sensitivity (V/m/S)
X	1.15	6.5097	43.96	37.99	26.32
Y	1.06	5.9315	35.40	40.52	24.68
Z	1.29	6.5128	22.86	37.43	26.72

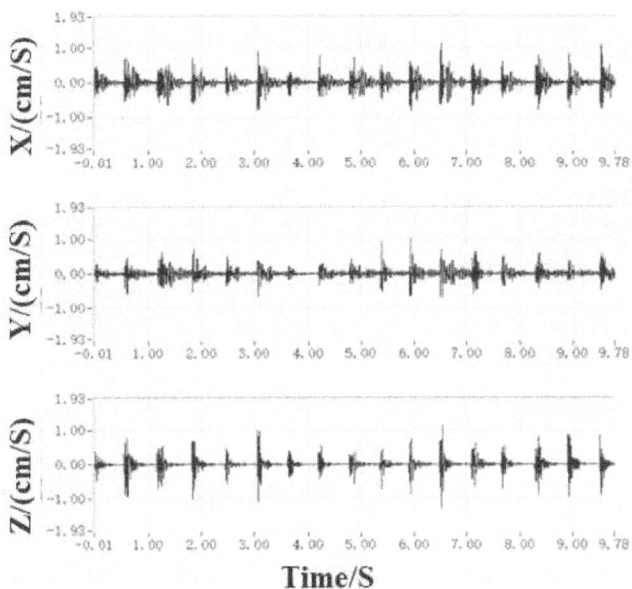

Fig.4.52 Blasting velocity in the X, Y and Z directions at point 2

The monitoring data of measuring point 3 are as follows:

Table 4.7 Data of measuring point 3

Chanel	Vib. Max (cm/S)	Time at max (S)	Main Freq. (Hz)	Range (cm/S)	Sensitivity (V/m/S)
X	0.93	8.4839	13.70	39.67	25.21
Y	0.98	8.1280	13.75	39.84	25.10
Z	1.12	8.0787	15.50	38.67	25.86

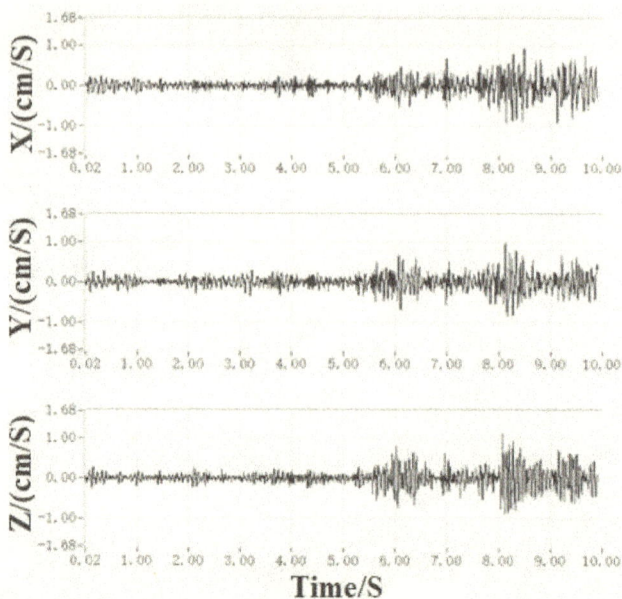

Fig.4.53 Blasting velocities in the X, Y and Z directions at point 3

The monitoring data of measuring point 4 are as follows:

Table.4.8 Data of measuring point 4

Chanel	Vib. Max (cm/S)	Time at max (S)	Main Freq. (Hz)	Range (cm/S)	Sensitivity (V/m/S)
X	0.70	2.4240	14.60	37.99	26.32
Y	0.68	0.0866	21.86	40.52	24.68
Z	0.96	0.0340	40.82	37.43	26.72

116

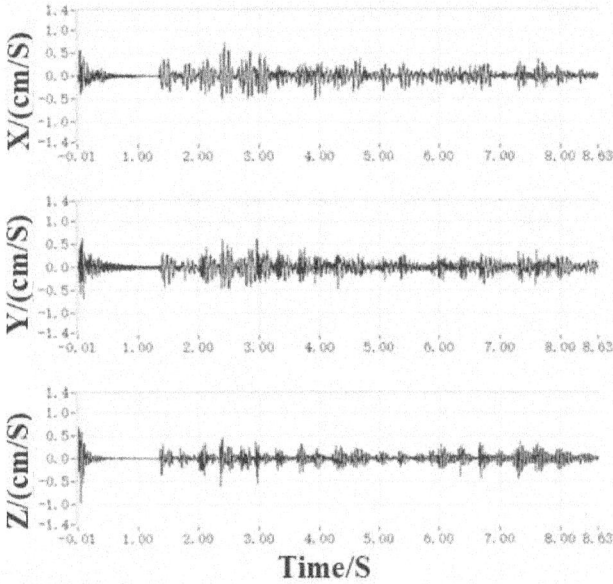

Fig.4.54 Blasting velocities in the X, Y and Z directions at point 4

The monitoring data of measuring point 5 are as follows:

Table.4.9 Data of measuring point 5

Chanel	Vib. Max (cm/S)	Time at max (S)	Main Freq. (Hz)	Range (cm/S)	Sensitivity (V/m/S)
X	0.62	0.1010	20.10	37.99	26.32
Y	0.59	0.7533	14.81	40.52	24.68
Z	0.79	0.2205	20.10	37.43	26.72

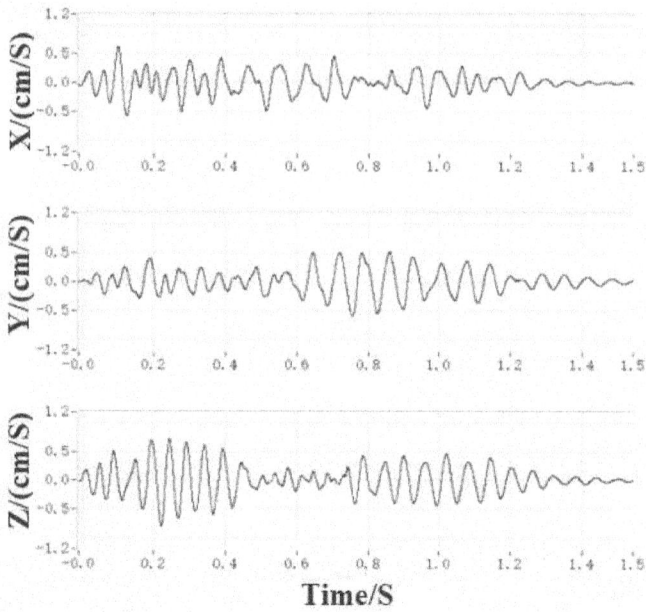

Fig.4.55 Blasting velocities in the X, Y and Z directions at point 5

The monitoring data of measuring point 6 are as follows:

Table.4.10 Data of measuring point 6

Chanel	Vib. Max (cm/S)	Time at max (S)	Main Freq. (Hz)	Range (cm/S)	Sensitivity (V/m/S)
X	0.24	0.6996	17.86	39.67	25.21
Y	0.47	0.6597	16.06	39.84	25.10
Z	0.23	0.1277	18.26	38.67	25.86

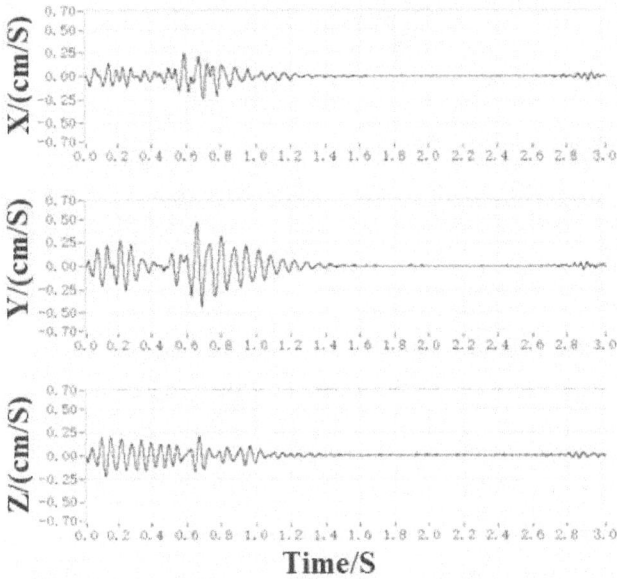

Fig.4.56 Blasting velocities in the X, Y and Z directions at point 6

3. Comparative analysis of theoretical calculation and monitoring results

According to the data in the above figures and tables, in the process of comparing the blasting vibration velocity and the main frequency of blasting vibration by using the proportional distance, the maximum blasting vibration velocities in the horizontal radial, horizontal tangential and vertical directions of blasting seismic waves recorded at each monitoring point are taken to obtain the relationship between the peak particle vibration velocity and the proportional distance, as shown in Figure 4.57.

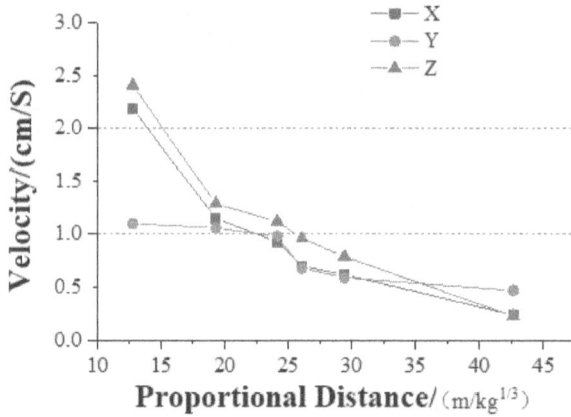

Fig.4.57 Variation of maximum vibration velocity

It can be seen from Figure 4.57, the peak vibration velocities of particles in the X, Yand Z directions decrease gradually with the proportional distance, but the attenuation amplitude in each direction is different. When the proportional distance is in the range of 10 to 20, the blasting vibration velocity in the Z direction is the maximum, which is 2.41cm/S; the maximum blasting vibration velocity in the X direction is 2.19cm/S; the peak particle blasting velocity in the Y direction is the minimum, which is 1.1cm/S. The peak particle vibration velocities in the Z and Y direction have a greater change than those in the X direction. With the increase of proportional distance, the blasting vibration velocities in all directions continue to attenuate. Because the blasting vibration velocity in the Y and Z directions decreases rapidly, when the proportional distance is greater than 25 and when $\bar{R} = 42.62$, the blasting vibration velocity in the Y direction reaches the maximum value of 0.47cm/S; the blasting vibration velocity in the Z direction reaches the

120

minimum value of 0.223cm/S; and the peak attenuation of blasting vibration in all directions tends to stabilize.

By comparing with the numerical simulation results in the previous chapter, it can be found that whether in numerical simulations or in practical engineering, the peak particle vibration velocity of blasting vibration in the Z direction in the far blasting vibration area is greater than that in the X and Y directions, and the attenuation amplitude of blasting vibration velocity in the Z direction is the largest, followed by that in the X direction. The variation of blasting vibration velocity in the Y direction with proportional distance is smaller than that in the X and Z directions. Because the millisecond blasting technology is used in the actual project, and the centralized blasting state is simulated in the numerical calculation, there are differences in the velocity time-history curves. The velocity time-history curve in the numerical calculation can more intuitively and completely see the seismic wave vibration propagation characteristics generated by a blasting vibration. In the engineering example, the vibration propagation characteristics of multiple blasts are monitored at the same monitoring point. The proportional distance is introduced in this study according the amount of explosive and distance from the blasting center. The maximum vibration velocities at different monitoring points are taken for analysis, making the vibration data at different monitoring points more comparable. However, according to the recorded data, it can be seen that each section of blasting vibration at the same monitoring position will produce a certain degree of superposition effect.

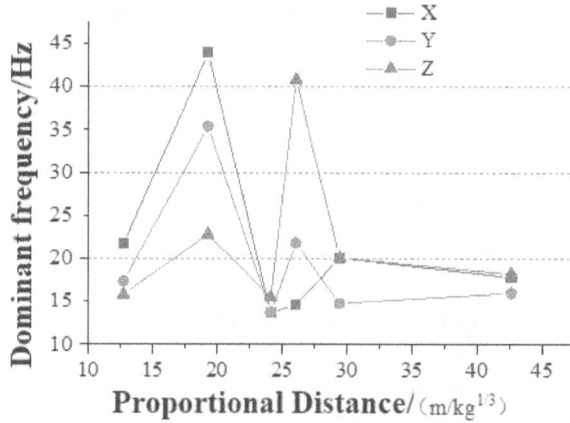

Fig.4.58 Variation curve of main blasting vibration frequency

The relationship between the main frequency of blasting vibration recorded by monitoring and the proportional distance is shown in Figure 4.58. Under the action of blasting, the stress wave acting on the rock produces the vibration effect in the X, Y and Z directions. It can be clearly seen from the above figure that when the proportional distance is within the range of 15 to 20, the main frequencies of blasting vibration in all directions have a numerical fluctuation, in which the main frequency of blasting vibration in the X direction is the largest compared with that in the Y and Z directions, reaching 43.96Hz. Afterwards, the main frequency attenuates sharply. When the proportional distance is within the range of 25 to 30, the main frequency of monitoring points fluctuates again. The main frequency of blasting vibration in the Z direction has the largest change range and finally tends to be stabilize. Compared with the numerical simulation results in the previous chapter, it can be concluded that during the propagation of seismic waves, the

122

main frequency of blasting vibration suddenly changes in the middle and far blasting vibration areas. However, in the actual monitoring data, due to the uneven medium in the rock mass, the different characteristics of the absorbed blasting seismic wave energy, and the unstable blasting vibration main frequency caused by the reflection superposition of the seismic waves in the process of propagation, the main frequency of blasting vibration more unstable. Therefore, compared with the numerical simulation, the blasting vibration frequency in the actual monitoring fluctuates more significantly as the proportional distance changes.

To better apply the findings to practical engineering and explore the general propagation law of blasting vibration, this chapter introduces the concept of proportional distance based on the amount of explosive and the distance from the blasting centers. The attenuation laws of blasting vibration velocity at measuring points #1~#8 are analyzed. MATLAB is also adopted to carry out Fourier transformation to convert the time-domain signals into the frequency-domain signals. By comparing them with actual engineering, the vibration frequencies are studied. The main results are as follows.

(1) For the blasting vibration velocity in the horizontal radial, horizontal tangential and vertical directions of measuring points #1~#8, this chapter divides the eight measuring points into two parts for comparative explorations. Through the analysis, it is found that the attenuation waveform of blasting vibration velocity in the X direction at each measuring point is similar and the vibration velocity decreases continuously with the increase of time. The

vibration velocity attenuation waveforms at measuring points #1~#4 in the Y and Z directions are more regular than those at position #5~#8. When at position #5~#8, the vibration states in the Y and Z directions are characterized by small fluctuations in vibration velocity in the early stage, but with the change of time, there is no obvious change in the amplitude of vibration velocity. Instead, the periods of vibration velocity fluctuations increase, with even up-and-down oscillations. At measuring points #7 and #8, the time taken to reach the peak particle vibration velocity is approximately 60MS.

(2) In this chapter, the maximum and minimum values of vibration velocity obtained by numerical simulation are analyzed. When the proportional distance is 1, the absolute value difference between the maximum and minimum values of blasting vibration velocity in the X direction is 0.796659m/S, followed by that in the Z and Y directions. With the increase of proportional distance, the maximum values of velocity in the X, Y and Z directions show a rapid downward trend at measuring point #2, with velocity change amplitudes of 63.7%, 72.4% and 74.3% respectively. The decrease amplitudes of the minimum values of blasting vibration (taking absolute values) are 37.3%, 20% and 90.8% respectively. When the proportional distance is greater than 30, the maximum and minimum values of blasting vibration in all directions are not monotonic attenuation but fluctuate to a certain degree.

(3) Based on the variation of peak particle vibration velocities with proportional distance from numerical simulation and theoretical calculation,

124

this chapter classifies the proportional distance of 1~5 as the near blasting vibration area, the proportional distance of 5~10 as the middle blasting vibration area, and the proportional distance greater than 10 as the far blasting vibration area.

(4) MATLAB is employed to perform Fourier transform on the blasting vibration velocities to obtain the blasting vibration spectrum of each measuring point. In the near blasting vibration area, the blasting vibration spectrum in the X direction has the widest distribution range, and each direction shows the characteristics of multi peak and wide distribution of blasting vibration frequency. When the proportional distance is 5~10 (i.e. the middle area of blasting vibration), the blasting vibration spectrum gradually transits from multiple dominant frequencies to a single dominant frequency, and the spectrum distribution range gradually narrows. When it is located in the far area of blasting vibration, the spectrum distribution is dominated by low-frequency components. At the same time, the main frequency of blasting vibration at each measuring point was compared, and it was found that the main frequency fluctuated suddenly in the process of attenuation with proportional distance. The first fluctuation of the main frequency of blasting vibration in X, Y and Z directions decreased by 51.43%, 50% and 27.78%, respectively, and the second fluctuation attenuation was 47.62%, 46.67% and 46.43%.

Furthermore, through the analysis of the monitoring point data of engineering examples, the variation laws of blasting vibration amplitude and dominant frequency with the proportional distance were found:

(1) Through the actual engineering monitoring data, it can be found that the variation range of particle peak vibration velocity in each direction of the monitoring point with the proportional distance from large to small is: Z direction, X direction, Y direction. When the proportional distance is in the range of 10 to 20, the maximum vibration velocity in Z direction is 2.41cm/S, the maximum vibration velocity in X direction is 2.19cm/S, and the maximum vibration velocity in Y direction is 1.1cm/S, which is 45.64% of the blasting vibration peak in Z direction; However, when the proportional distance is in the range of 35 to 45, the maximum vibration velocity in Y direction is greater than that in X and Z directions due to the slow attenuation of blasting vibration peak in Y direction.

(2) By analyzing the monitoring data and numerical simulation results of engineering examples, it is found that in the process of seismic wave propagation, the dominant frequency of vibration does not decay monotonously with the proportional distance, but will fluctuate in the middle zone of blasting vibration. This conclusion is consistent with the numerical simulation results, and due to the uneven distribution of the actual engineering rock mass medium, the main frequency fluctuation is more obvious than the numerical simulation. When the proportional distance is between 15 and 20, the main frequency fluctuation amplitude of blasting vibration is in the following order: X direction, Y direction, Z direction, but when the proportional distance is between 25 and 30, the main frequency fluctuation of

126

blasting vibration in the Z direction is larger, while the fluctuations in the Y direction and X direction are smaller.

(3) The results of monitoring data analysis in practical engineering verify the conclusion that the blasting vibration characteristics continuously attenuate with proportional distance in rock blasting. At the same time, the attenuation amplitude of blasting vibration velocity in Z direction is larger than that in other directions, and the dominant frequency of blasting vibration will fluctuate to a certain extent in the middle and far regions of blasting vibration in the attenuation process. This conclusion is consistent with the numerical simulation results in this study, proving that the research method is reasonable and the research results are valid.

5 Blasting Vibration Field with Defective Inclusions

‒‒‒‒‒‒‒‒ ━◑ ➤➤ ◉ ◆ ✳ ◆ ◉ ◀◀ ◐━ ‒‒‒‒‒‒‒‒

5.1 Simulation with Defective Geological Inclusion

When the ground vibration intensity exceeds the critical threshold that structures can withstand, buildings and structures on the ground are likely to be damaged. Therefore, to control the blasting hazards, it is necessary to control the blasting vibration intensity experienced by structures. Many factors influence the blasting vibration intensity, and one of the most commonly-used indicators to judge the blasting vibration intensity is the distance from the explosion center, that is, the distance from the location to the explosion point [84]. Thus, after completing numerical simulation calculations, the monitoring points with appropriate distance should be selected for data verification and analysis.

According to equation (4.5) and the established model, seven measuring points #1~#7 are selected along the horizontal and radial direction of the concentrated explosive, with proportional distances of $1m/kg^{1/3}$, $3m/kg^{1/3}$, $4m/kg^{1/3}$, $7m/kg^{1/3}$, $10m/kg^{1/3}$, $15m/kg^{1/3}$ and $20m/kg^{1/3}$ respectively, as shown in Figure 5.1. The time-history curves of blasting vibration velocity at these seven monitoring points are extracted and analyzed to study the influence of different ground inclusions on the propagation law of blasting seismic wave.

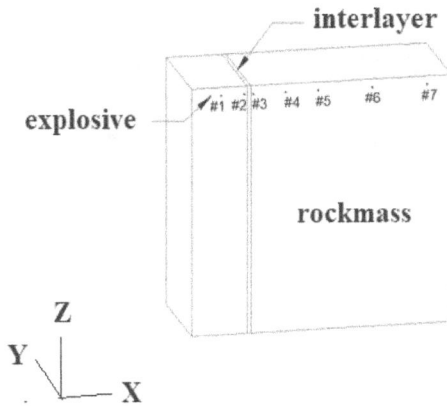

Fig.5.1 Location map of measurement points

After extracting the simulation data from measuring points #1~#7 in the model, an analysis is conducted on the velocity time-history curves in the horizontal radial, horizontal tangential and vertical directions, i.e. X, Y and Z directions. The analysis focuses on the variation laws of particle blasting vibration velocity under the blasting action at proportional distances of shows that the proportional distance $\overline{R}=1$ m/kg$^{1/3}$, $\overline{R}=3$ m/kg$^{1/3}$, $\overline{R}=4$ m/kg$^{1/3}$, $\overline{R}=7$ m/kg$^{1/3}$, $\overline{R}=10$ m/kg$^{1/3}$, $\overline{R}=15$ m/kg$^{1/3}$ and $\overline{R}=20$ m/kg$^{1/3}$.

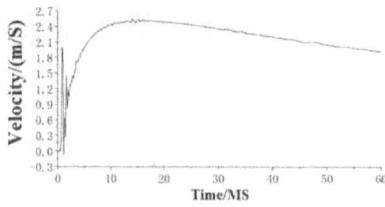

(a) granite interlayer

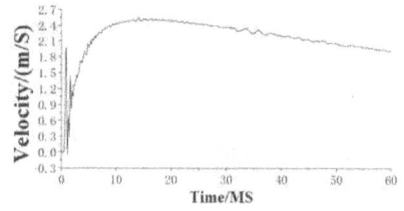

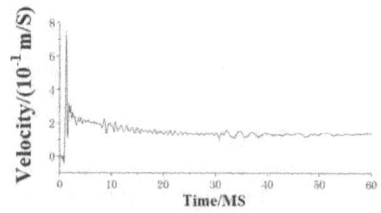

(b) limestone interlayer

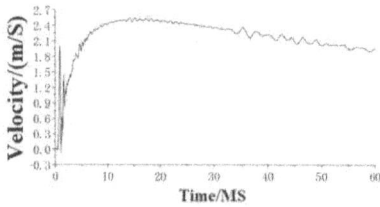

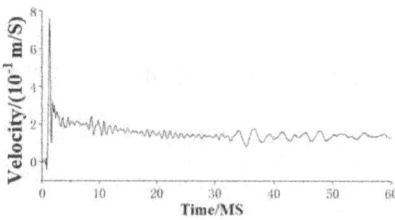

(c) crushed stone interlayer

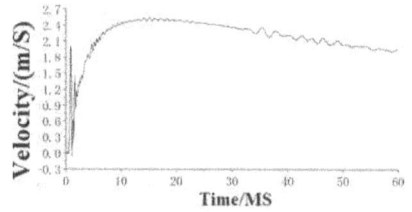

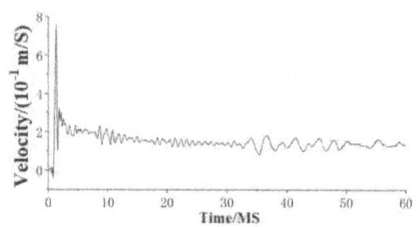

(d) hard clay interlayer

Fig.5.2 Velocity time-history curves in X direction at #1

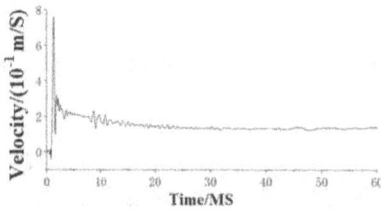

(a) granite interlayer

(b) limestone interlayer

(c) crushed stone interlayer

(d) hard clay interlayer

Fig.5.3 Velocity time-history curves in Y direction at #1

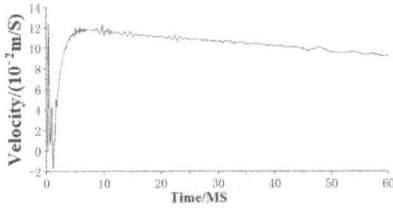

(a) granite interlayer   (b) limestone interlayer

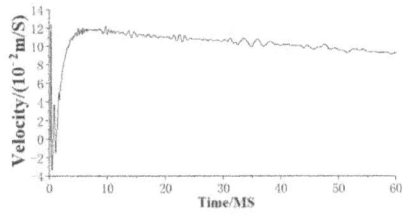

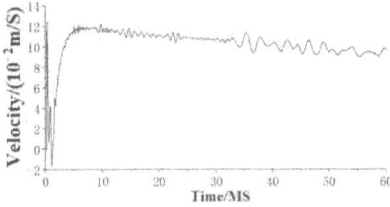

(c) crushed stone interlayer   (d) hard clay interlayer

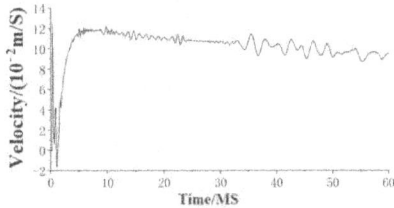

Fig.5.4 Velocity time-history curves in Z direction at #1

Based on Figures 5.2 to 5.4, an analysis is performed on the velocity time-history curves in the X, Y and Z directions at the position with a scaled distance of 1 $m/kg^{1/3}$, yielding the following results:

(1) Under the action of blasting, the vibration velocity at measuring point #1 fluctuates sharply. The velocity time-history curves in the X, Y and Z directions oscillate violently within the time range of 0~3MS. The minimum value of blasting vibration velocity in the whole blasting process is obtained, and the maximum value of blasting vibration velocity in Y and Z directions is also obtained within this time period, while the maximum value of blasting vibration in the X direction is obtained in about 15MS.

(2) Because measuring point #1 is separated from the strata by a certain distance, the trends of velocity time-history curves of measuring point #1 in

the X, Y and Z directions is roughly the same across different stratum materials. The influence of the strata on the velocity time-history curves is mainly reflected in the varying oscillation amplitudes of the curves after 30MS.

(3) After the oscillation period of 0-3MS, the velocity time-history curve in the X direction continues to rise under the action of subsequent blasting stress waves and begins to decline slowly after reaching the maximum value in about 15MS. When the stratum is composed of granite, dolomitic limestone, gravelly soil and hard clay, the maximum blasting vibration velocity in the X direction is 2.5376 m/S, 2.5369m/S, 2.5437m/S and 2.5473m/S respectively, and the minimum is -5.7632×10^{-2}m/S, -5.4082×10^{-2}m/S, -5.7631×10^{-2}m/S and -5.7632×10^{-2}m/S respectively. The maximum blasting vibration velocity difference in the X direction is not more than 1%, and the minimum blasting vibration velocity in dolomitic limestone is about 94% of that in other materials. Because 25kg explosive is used and point #1 is too close to the explosion source, the vibration of the velocity time-history curve in the X direction is mainly positive vibration, that is, vibration away from the explosion point.

(4) In the oscillation period of 0~3MS, the velocity time-history curve in the Y direction achieves the maximum and minimum values. After reaching the maximum value, the oscillation amplitude of the velocity time-history curve begins to decrease, and hovers in the range of 10^{-1}m/S to 2×10^{-1}m/S after 30MS. When the stratum is composed of granite, dolomitic limestone,

gravelly soil and hard clay, the maximum blasting vibration velocity in the Y direction is 7.5928×10^{-1}m/S, 7.5187×10^{-1}m/S, 7.5928×10^{-1}m/S and 7.5928×10^{-1}m/S respectively, and minimum values are -4.1039×10^{-2}m/S, -4.1406×10^{-2}m/S, -4.1038×10^{-2}m/S and -4.1038×10^{-2}m/S respectively. The maximum blasting vibration velocity difference in the Y direction is not more than 1%, and the minimum blasting vibration velocity difference is not more than 1%. Because 25kg explosive is used and point #1 is too close to the explosion source, the vibration of velocity time-history curve in the Y direction is mainly positive vibration, that is, vibration away from the explosion point.

(5) In the oscillation period of 0-3MS, the velocity time-history curve in the Z direction achieves the maximum and minimum values. Under the action of the subsequent blasting stress waves, the velocity time-history curve continues to rise and hovers in the range of 8×10^{-2}m/S to 11×10^{-2}m/S after 30MS. When the stratum is composed of granite, dolomitic limestone, gravelly soil and hard clay, the maximum blasting vibration velocity in the Z direction is 1.2376×10^{-1}m/S, 1.2413×10^{-1}m/S, 1.2376×10^{-1}m/S and 1.2376×10^{-1}m/S respectively, and the minimum values are -1.6618×10^{-2}m/S, -3.3382×10^{-2}m/S, -1.6618×10^{-2}m/S and -1.6618×10^{-2}m/S respectively. Therefore, the difference of the maximum blasting vibration velocity in the Z direction is no more than 1%. Only when the stratum is composed of dolomitic limestone, the minimum blasting vibration velocity is about twice that of other materials. Because 25kg explosive is used and point #1 is too close to the

133

explosion source, the vibration of the velocity time-history curve of the vibration in the Z direction is mainly positive vibration, that is, the vibration away from the explosion point.

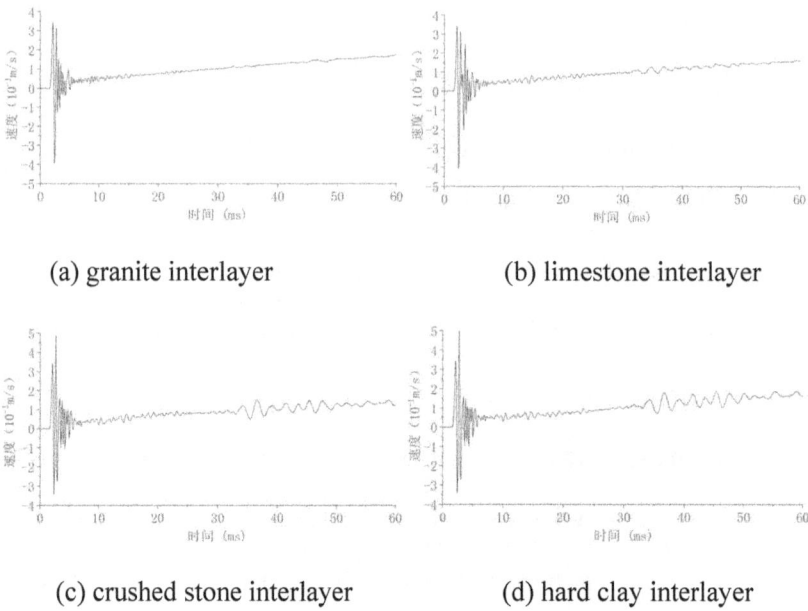

(a) granite interlayer   (b) limestone interlayer

(c) crushed stone interlayer   (d) hard clay interlayer

Fig.5.5 Velocity time-history curves in X direction at #2

Based on Figures 5.5-5.7, an analysis is conducted on the velocity time-history curves in the X, Y and Z directions at the position with a proportional distance of 43 m/kg$^{1/3}$, yielding the following results:

(1) Due to the special location of measuring point #2 between the explosion source and the stratum and in close proximity to the stratum, measuring point #2 is affected by the stratum's reflected wave effect, which arises from the different grids set between the stratum and the rock mass and the contact mode. Because only 60MS of calculation time is set, the total

calculation time is too short, leading to an upward trend in the calculation time within 0~60MS in the time-history curve of measuring point #2. As measuring point #2 has not crossed the stratum, the trend of velocity time-history curves in the X, Y and Z directions are roughly the same. The influence of stratum on velocity time-history curves is reflected in the varying oscillation amplitudes of velocity time-history curves after 35MS.

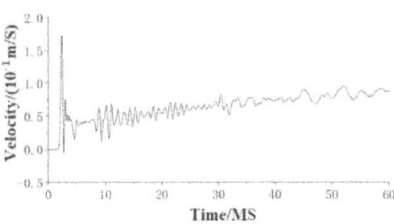

(a) granite interlayer

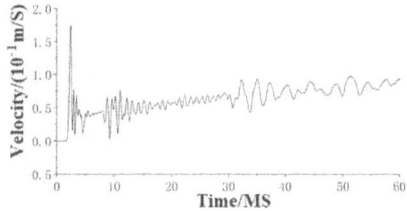

(b) limestone interlayer

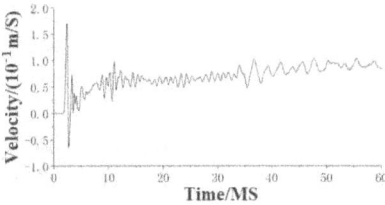

(c) crushed stone interlayer

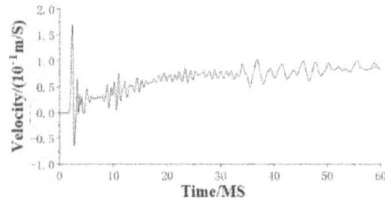

(d) hard clay interlayer

Fig.5.6 Velocity time-history curves in Y direction at #2

(2) Under the action of blasting, the vibration velocity at measuring point #2 fluctuates sharply. The velocity time-history curves in the X and Y directions oscillate violently in 2~5MS. In most cases, both the maximum and minimum values of blasting vibration velocity in the whole blasting process are obtained. The velocity time-history curve in the Z direction oscillates

violently within 2~5MS, and the minimum value of blasting vibration velocity is obtained in this process, but the oscillation of velocity time-history curve does not stop until 35MS later.

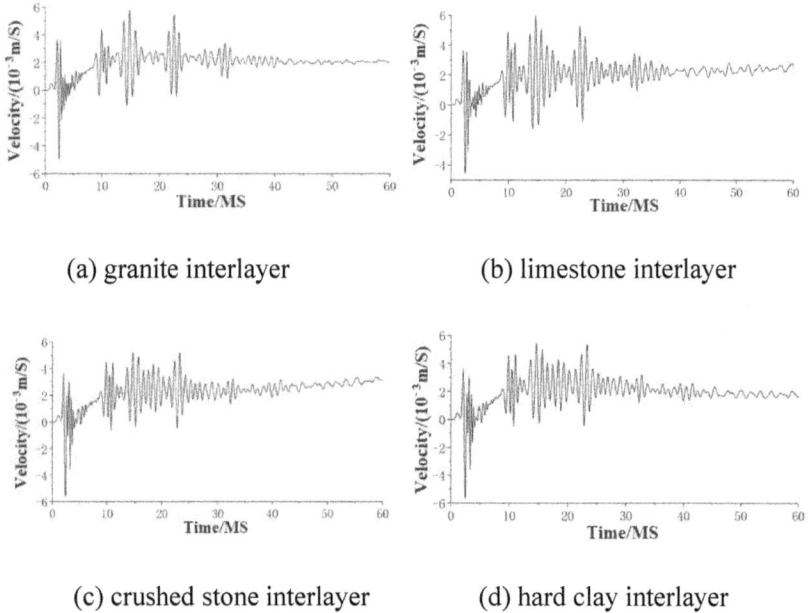

(a) granite interlayer   (b) limestone interlayer

(c) crushed stone interlayer   (d) hard clay interlayer

Fig.5.7 Velocity time-history curves in Z direction at #2

(3) During the oscillation period of 2~5MS, the velocity time-history curve in the X direction reaches both the maximum and minimum values of the blasting vibration velocity in the whole blasting process. Immediately afterwards, the curve tends to stabilize while continuing to rise slowly under the action of subsequent blasting stress waves. The value is about 1.93×10^{-1} m/S at 60MS. When the stratum is composed of granite, dolomitic limestone, gravelly soil and hard clay, the maximum blasting vibration velocity in the X

direction is 3.4460×10^{-1}m/S, 3.3782×10^{-1}m/S, 4.8725×10^{-1}m/S and 4.9840×10^{-1}m/S respectively, and the minimum values are -3.9361×10^{-1}m/S, -4.0479×10^{-1}m/S, -3.4388×10^{-1}m/S and -3.4205×10^{-1}m/S respectively. Therefore, a measuring point #2, the maximum and minimum blasting vibration velocities in the X direction are of the same order of magnitude, with little difference in their numerical values.

(4) In the oscillation period of 2~5MS, the velocity time-history curve in the Y direction reaches both the maximum and minimum values except when the stratum is composed of dolomitic limestone. After reaching the maximum value, the velocity time-history curve become smaller in both oscillation amplitude and frequency. After 30MS, the oscillation amplitude increases, but the oscillation frequency continues to decrease. When the stratum is composed of dolomitic limestone, the maximum value of the velocity-time-history curve in the Y direction is obtained in the oscillation period of 2~5MS, and the minimum value is obtained in about 9MS. When the stratum is composed of granite, dolomitic limestone, gravelly soil and hard clay, the maximum blasting vibration velocity in the Y direction is 1.7297×10^{-1}m/S, 1.7525×10^{-1}m/S, 1.6986×10^{-1}m/S and 1.6981×10^{-1}m/S respectively, and the minimum values are -5.5561×10^{-3}m/S, -1.1742×10^{-4}m/S, -6.3901×10^{-2}m/S and -6.6351×10^{-2}m/S respectively. The vibration of the time-history curve of the vibration velocity in the Y direction at measuring point #2 is mainly positive vibration, that is, the vibration in the direction away from the explosion point. The maximum value of the blasting vibration velocity is of the same order in

137

magnitude with little difference, and its maximum values are 31.13 times, 1492.57 times, 2.66 times and 2.56 times of its minimum value, respectively. It can be seen that the stratum material has a great influence on the minimum vibration velocity in the Y direction at measuring point #2. The harder the stratum, the smaller the minimum vibration velocity and the greater the difference between the minimum vibration velocity and the corresponding maximum vibration velocity. The softer the stratum, the greater the minimum vibration velocity and the smaller the difference between the minimum vibration velocity and the corresponding maximum vibration velocity.

(5) In the oscillation period of 2~5MS, the velocity time-history curve in Z direction oscillates violently and achieves the minimum value of blasting vibration velocity. Under the action of subsequent blasting stress waves, the velocity time-history curve oscillates continuously and shows an upward trend, and achieves the maximum value in about 14MS. When the stratum is composed of granite, dolomitic limestone, gravelly soil and hard clay, the maximum blasting vibration velocity in the Z direction is 5.7932×10^{-3}m/S, 6.1274×10^{-3}m/S, 5.2415×10^{-3}m/S and 5.4410×10^{-3}m/S respectively, and the minimum values are -4.9504×10^{-3}m/S, -4.5528×10^{-3}m/S, -5.6036×10^{-3}m/S and -5.6393×10^{-3}m/S respectively. Thus, the maximum blasting vibration velocity in the Z direction is of the same order in magnitude as the difference between the minimum and the maximum values is within 20%. Within 2~5MS, the vibration of the velocity time-history curve of the vibration in the Z direction at measuring point #2 is mainly negative vibration, that is, the

138

vibration in the direction close to the explosion point. After 5MS, the vibration of the velocity time-history curve is mainly positive vibration, that is, the vibration away from the explosion point.

(6) The vibration velocity of measuring point #2 in the Z direction shows multiple peaks and troughs in the velocity time-history curve. An analysis reveals that this phenomenon occurs because measuring point #2 is close to the stratum and is thus affected by the reflection and transmission effect of seismic waves. When blasting seismic waves touch the stratum and free surface, reflection and transmission occur, resulting in new seismic waves. These new waves then overlap

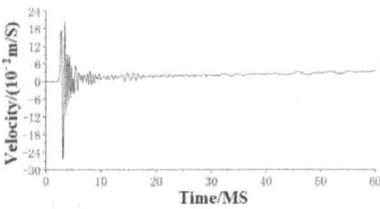

(a) granite interlayer   (b) limestone interlayer

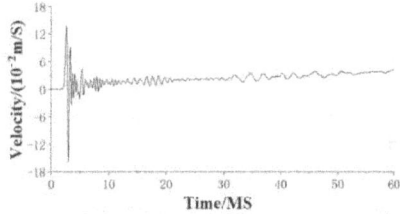

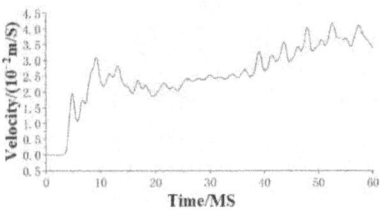

(c) crushed stone interlayer   (d) hard clay interlayer

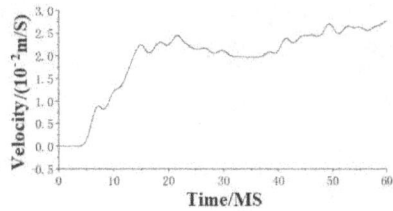

Fig.5.8 Velocity time-history curves in X direction at #3

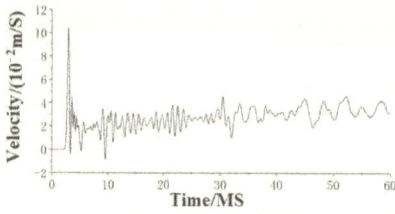

(a) granite interlayer

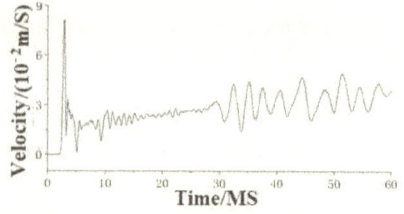

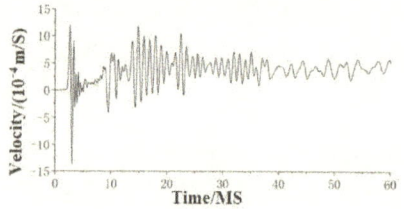

(b) limestone interlayer

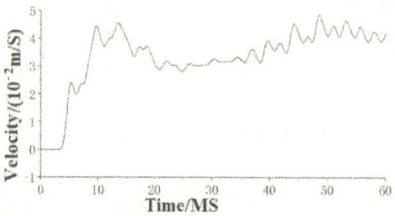

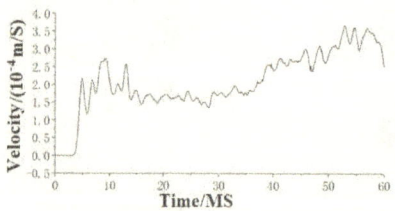

(c) crushed stone interlayer

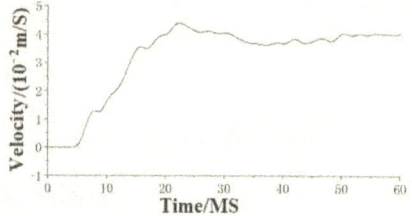

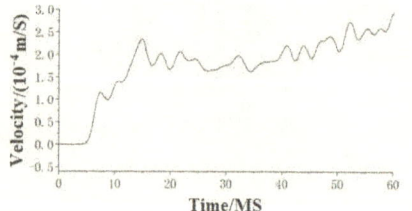

(d) hard clay interlayer

Fig.5.9 Velocity time-history curves in Y direction at #3

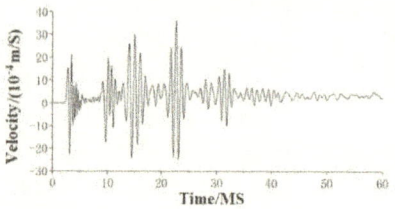

(a) granite interlayer

(b) limestone interlayer

(c) crushed stone interlayer

(d) hard clay interlayer

Fig.5.10 Velocity time-history curves in Z direction at #3

Based on Figures 5.8-5.10, an analysis of the velocity time-history curves in the X, Y and Z directions at the position with a proportional distance of $4m/kg^{1/3}$ yields the following results:

(1) For the close location of measuring point #3 to the stratum, the time-history curve of measuring point #3 shows an upward trend due to the influence of the reflection wave effect of the stratum. Under the action of blasting, the hard stratum has little effect on the vibration waveform of measuring point #3, which is mainly reflected in the different oscillation amplitude of the velocity time-history curve and the different time of the maximum and minimum values of the vibration velocity after 30MS. However, the weak stratum greatly changes the vibration waveform of the velocity time-history curve of measuring point #3. Compared with the vibration waveform of the hard stratum, the vibration frequency of the seismic wave after passing through the weak stratum is significantly reduced.

(2) For the velocity time-history curve in the X direction, when the stratum is composed of granite and dolomitic limestone, the velocity time-history curve oscillates violently in 3~6MS, and the maximum and minimum values of the blasting vibration velocity in the whole blasting process are obtained. Afterwards, the vibration amplitude and vibration frequency of the velocity time-history curve decrease. The maximum values of the vibration velocity during blasting are $2.0506\times10^{-1}m/S$ and $1.3843\times10^{-1}m/S$ respectively and the minimum values are $-2.6217\times10^{-1}m/S$ and $-1.5654\times10^{-1}m/S$ respectively. The maximum and minimum values of the vibration velocity

141

when the stratum is composed of dolomitic limestone are 67.51% and 59.71%

of the values when the stratum is composed of granite. When the stratum is

composed of gravel soil and hard clay, the velocity time-history curve shows

a trend of first rising, then falling and finally rising within 0~60MS. The

maximum values of vibration velocity are obtained within 48~53MS, which

are 4.1840×10^{-2}m/S and 2.7738×10^{-2}m/S respectively, 20.40% and 13.53%

of the maximum values of vibration velocity when the stratum is composed

of granite. Because the velocity time-history curves of both cases are basically

upward trend, the minimum values of vibration velocity of both are 0m/S.

(3) For the velocity time-history curve in the Y direction, when the

stratum is composed of granite and dolomitic limestone, the velocity time-

history curve oscillates violently within 3~5MS and the maximum value of

the blasting vibration velocity in the whole blasting process is obtained.

Afterwards, the oscillation amplitude of the velocity time-history curve and

the oscillation frequency decrease first and then stabilize gradually. After

30MS, the oscillation amplitude of the velocity time-history curve of both

increases while the oscillation frequency continues to decrease. The maximum

vibration velocities during blasting are 1.0407×10^{-1}m/S and 8.1452×10^{-2}m/S

respectively and the minimum vibration velocities are -8.5205×10^{-3}m/S and $-$

4.1568×10^{-6}m/S respectively. The maximum and minimum vibration

velocities when the formation is composed of dolomitic limestone are 78.26%

and 0.05% of those when the stratum is composed of granite, respectively.

When the stratum is composed of gravel soil and hard clay, the oscillation frequency of the velocity time-history curve is much lower than that when the stratum is composed of granite and dolomitic limestone. Due to the short total calculation time because only 60MS is set for calculation, the time-history curve generally shows a trend of first rising, then falling, and then rising within 0~60MS. The maximum vibration velocities of the two curves are 4.8803×10^{-2}m/S and 4.4051×10^{-2}m/S respectively, which are 46.89% and 42.33% of the maximum vibration velocities when the stratum is composed of granite. Because the velocity time-history curves of the two strata exhibit a basically upward trend, the minimum vibration velocities of the two strata are -3.5510×10^{-8}m/S and -3.5510×10^{-8}m/S respectively. -2.4852×10^{-10}m/S, which is less than one thousandth of the minimum value of granite vibration velocity.

(4) For the velocity time-history curve in the Z direction, when the stratum is composed of granite, the velocity time-history curve oscillates violently within 3~5MS and the oscillation amplitude decreases sharply within 5~7MS. Afterwards, the oscillation frequency decreases while the oscillation amplitude increases. The velocity time-history curve after 33MS tends to stabilize gradually. The maximum value of vibration velocity during blasting is 3.6149×10^{-3}m/S, and the minimum value is -2.5124×10^{-3}m/S. When the stratum is composed of dolomitic limestone, the velocity time-history curve oscillates violently within 3~5MS, and the maximum and minimum values of the blasting vibration velocity in the whole blasting process are obtained. The oscillation amplitude decreases sharply within

143

5~8MS and then the oscillation frequency decreases while the oscillation amplitude increases. After 38MS, the velocity time-history curve tends to stabilize gradually. During blasting, the maximum value of the vibration velocity is 1.1914×10^{-3}m/S, and the minimum value is -1.3574×10^{-3}m/S. he maximum and minimum values of the vibration velocity when the stratum is composed of dolomitic limestone are 32.96% and 54.03% of the values when the stratum is composed of granite. The oscillation frequency of the velocity time-history curve when the stratum is composed of gravel soil and hard clay is much lower than that when the stratum is granite and dolomitic limestone. Within 0~60MS, the curve shows a trend of first rising, then falling, and then rising. The maximum vibration velocity is 3.6828×10^{-4}m/S and 2.9672×10^{-4}m/S, respectively, which are 10.19% and 24.90% of the maximum vibration velocity when the stratum is composed of granite. Because the velocity time-history curves of both are basically upward trend, the minimum vibration velocities of both are -5.7511×10^{-11}m/S and -3.8426×10^{-10}m/S respectively, which are less than 10 000% of the minimum vibration velocity when the stratum is composed of granite.

(5) The vibration velocity of measuring point #3 in the Z direction shows multiple peaks and troughs in the velocity time-history curve. An analysis reveals that this phenomenon occurs because measuring point #3 is close to the stratum and is thus affected by the reflection and transmission effect of seismic waves. When blasting seismic waves touch the stratum and free surfaces, reflection and transmission occur, resulting in new seismic waves.

These new waves then overlap with or counteract the seismic waves directly propagating from the explosion source to the measuring point. Therefore, there will be multiple peaks and troughs in the Z direction velocity time-history curve of the measuring point.

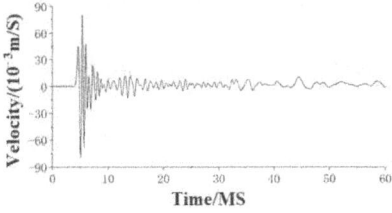

(a) granite interlayer (b) limestone interlayer

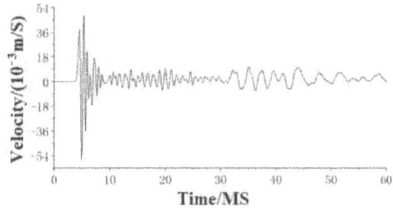

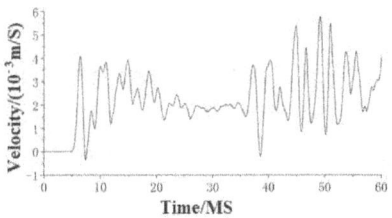

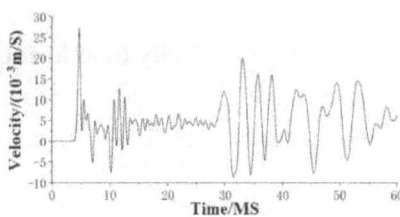

(c) crushed stone interlayer (d) hard clay interlayer

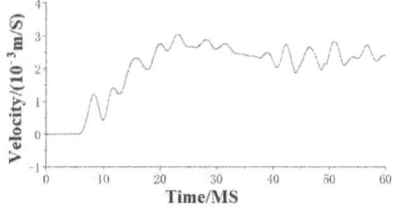

Fig.5.11 Velocity time-history curves in X direction at #4

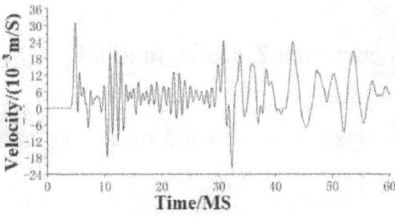

(a) granite interlayer (b) limestone interlayer

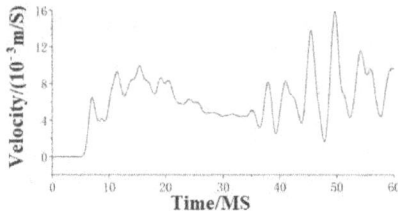

(c) crushed stone interlayer   (d) hard clay interlayer

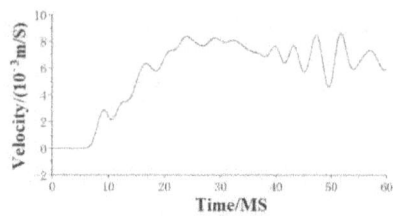

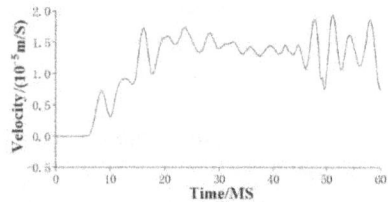

Fig.5.12 Velocity time-history curves in Y direction at #4

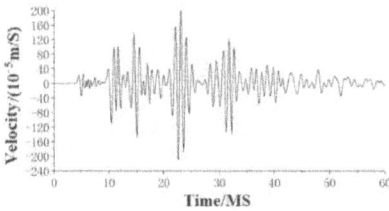

(a) granite interlayer   (b) limestone interlayer

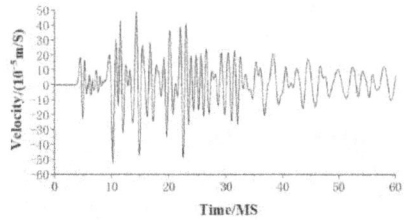

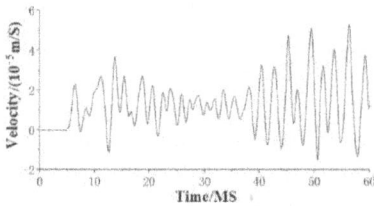

(c) crushed stone interlayer   (d) hard clay interlayer

Fig.5.13 Velocity time-history curves in Z direction at #4

Based on Figures 5.11-5.13, an analysis is conducted on the velocity time-history curves in the X, Y and Z directions at the position with a proportional distance of 43 m/kg$^{1/3}$, yielding the following results:

(1) Under the action of blasting, the hard stratum has little effect on the vibration waveform of measuring point #4 and the effect is mainly reflected

146

in the oscillation amplitude difference of the velocity time-history curve and the timing of the maximum and minimum values of the vibration velocity after 30MS. But the weak stratum greatly changes the vibration waveform of the velocity time-history curve of measuring point #4. Compared with the vibration waveform of the hard stratum, the vibration frequency of the seismic waves after passing through the weak stratum is greatly reduced.

(2) For the velocity time-history curve in the X direction, when the stratum is composed of granite and dolomitic limestone, the velocity time-history curve oscillates violently within 4~7MS and the maximum and minimum values of the blasting vibration velocity in the whole blasting process are obtained. Afterwards, both the oscillation amplitude and frequency of the velocity time-history curve decrease, but the oscillation amplitude rises while the oscillation frequency continues to decrease after 30 MS. The maximum values of the vibration velocity in the blasting vibration are 8.0241×10^{-2}m/S and 4.8515×10^{-2}m/S respectively and the minimum values of the vibration velocity are -7.8907×10^{-2}m/S and -5.6762×10^{-2}m/S respectively. When the stratum is composed of dolomitic limestone, the maximum and minimum values of the vibration velocity in the blasting vibration are 60.46% and 71.93% respectively of those when the stratum is composed of granite. When the stratum is composed of gravel soil and hard clay, the curve shows a trend of first rising and then falling within 0~60MS while the oscillation amplitude shows a trend of first rising, then falling and then rising again. The maximum vibration velocity in the blasting vibration of

147

the two strata are 5.7851×10^{-3}m/S and 3.0550×10^{-3}m/S respectively, which are 7.21% and 3.81% of the maximum vibration velocity when the stratum is composed of granite. Because the velocity time-history curves of the two strata basically show an upward trend, the minimum vibration velocity in the blasting vibration are -3.6502×10^{-4}m/S and -3.0880×10^{-14}m/S respectively, both of which are less than 11% of the minimum vibration velocity when the stratum is composed of granite.

(3) For the velocity time-history curve in the Y direction, when the stratum is composed of granite and dolomitic limestone, the velocity time-history curve directly reaches the maximum value of vibration velocity at about 4MS, and then the oscillation amplitude decreases gradually until it increases sharply again after 28MS. The oscillation frequency continues to decrease and the minimum value of vibration velocity is obtained at about 32MS. Afterwards, the oscillation amplitude decreases again with the passage of time. The maximum values of vibration velocity during blasting are 3.1001×10^{-2}m/S and 2.7272×10^{-2}m/S respectively, and the minimum values are -2.1769×10^{-2}m/S and -8.6353×10^{-2}m/S respectively. When the strata are composed of dolomitic limestone, the maximum and minimum values of vibration velocity are 87.97% and 39.67% respectively of those when the stratum is composed of granite. When the stratum is composed of gravel soil and hard clay, the oscillation frequency of the velocity time-history curve is much lower than that when the stratum is composed of granite and dolomitic limestone. The curve shows a trend of first rising and then falling within

148

0~60MS while the oscillation amplitude shows a trend of first increasing, then decreasing and then increasing as a whole. The maximum vibration velocity in the blasting vibration of the two strata are 1.5856×10^{-2}m/S and 8.6479×10^{-3}m/S respectively, which are 51.15% and 27.90% of the maximum vibration velocity when the stratum is composed of granite. Because the velocity time-history curves of the two strata are basically upward trend, the minimum values of vibration velocity in the blasting vibration are -5.3046×10^{-11}m/S and -1.1543×10^{-14}m/S respectively, which are less than one ten-thousandth of the minimum value of vibration velocity when the stratum is composed of granite.

(4) For the velocity time-history curve in the Z direction, when the stratum is composed of granite and dolomitic limestone, the oscillation amplitude of the velocity time- history curve increases sharply after 8MS, and then the maximum and minimum values of the vibration velocity in the whole blasting vibration are obtained. After 37MS, the oscillation frequency of the velocity time-history curve decreases significantly. The maximum values of the vibration velocity in the blasting vibration are 2.1118×10^{-3}m/S and 4.9254×10^{-4}m/S respectively, and the minimum values of the vibration velocity in the blasting vibration are -2.0986×10^{-3}m/S and -5.2751×10^{-4}m/S respectively. When the stratum is composed of dolomitic limestone, the maximum and minimum values of the vibration velocity are 23.32% and 25.14% respectively of the values when the stratum is composed of granite. When the stratum is composed of gravel soil and hard clay, the oscillation frequency of the velocity time-history curve is much lower than that when the

149

stratum is composed of granite and dolomitic limestone, showing a trend of first rising and then falling within 0~60MS while the oscillation amplitude shows a trend of first rising, then falling and then rising again as a whole. The maximum vibration velocity in the blasting vibration of the two strata are 5.2869×10^{-5}m/S and 1.9437×10^{-5}m/S respectively, which are 2.50% and 3.95% of the maximum vibration velocity when the stratum is composed of granite. Because the velocity time-history curves of the two strata basically takes on an upward trend, the minimum values of vibration velocity in the blasting vibration are -1.5491×10^{-5}m/S and 4941×10^{-13}m/S respectively, which are less than 1% of the minimum value of vibration velocity when the stratum is composed of granite.

(5) When the stratum is composed of granite and weathered dolomitic ash, the velocity time-history curve of measuring point #4 in the Z direction shows that there are multiple peaks and troughs. An analysis reveals that the seismic waves transmitted to the measuring point is less affected by the transmission and reflection effects of seismic waves because the measuring point is a certain distance away from the stratum. Therefore, the velocity time-history curve of measuring point #4 in the Z direction inherits the characteristics of the velocity time-history curve of measuring point #3 in the Z direction.

Based on Figures 5.14~5.16, an analysis is conducted on the velocity time-history curves in the X, Y and Z directions at the position with a proportional distance of 10 m/kg$^{1/3}$, yielding the following results:

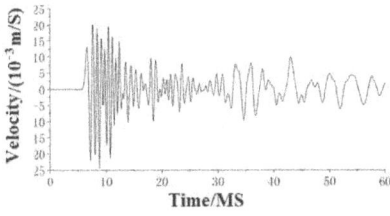

(a) granite interlayer   (b) limestone interlayer

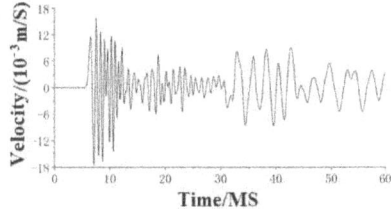

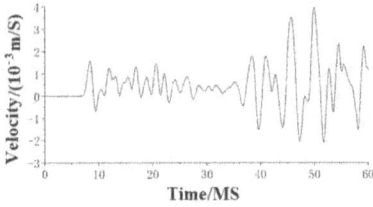

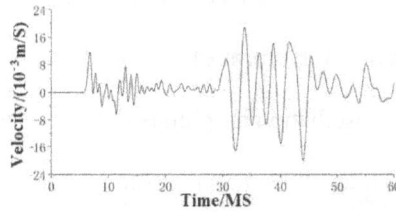

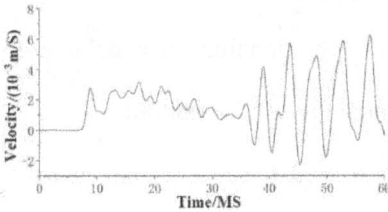

(c) crushed stone interlayer   (d) hard clay interlayer

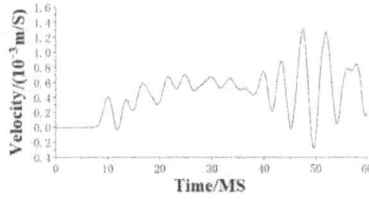

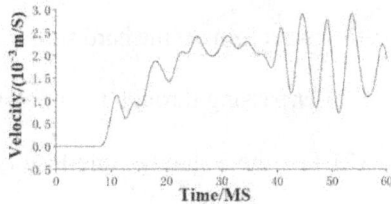

Fig.5.14 Velocity time-history curves in X direction at #5

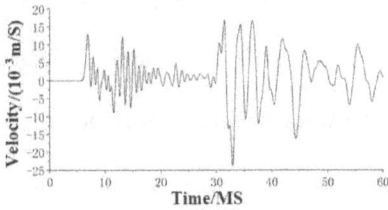

(a) granite interlayer   (b) limestone interlayer

(c) crushed stone interlayer   (d) hard clay interlayer

Fig.5.15 Velocity time-history curves in Y direction at #5

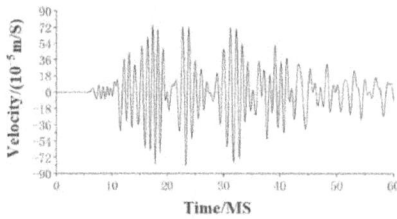

(a) granite interlayer   (b) limestone interlayer

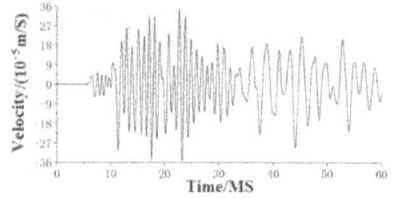

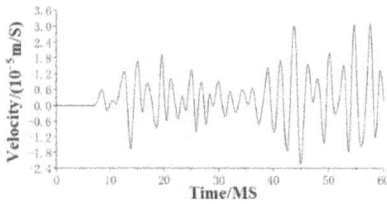

(c) crushed stone interlayer   (d) hard clay interlayer

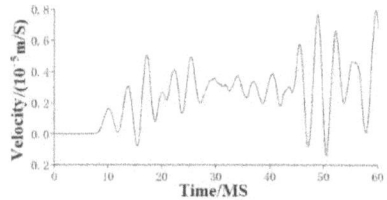

Fig.5.16 Velocity time-history curves in Z direction at #5

(1) Under the action of blasting, the hard stratum has little effect on the vibration waveform of measuring point #5 and the effect is mainly reflected in the oscillation amplitude difference of the velocity time-history curve and the timing of the maximum and minimum values of the vibration velocity after 30MS. But the weak stratum greatly changes the vibration waveform of the velocity time-history curve of measuring point #5. Compared with the vibration waveform of the hard stratum, the vibration frequency of the seismic waves after passing through the weak stratum is greatly reduced.

(2) For the velocity time-history curve in the X direction, when the stratum is composed of granite and dolomitic limestone, the velocity time-history curve oscillates violently in 6~11MS, and the maximum and minimum

values of the blasting vibration velocity in the whole blasting process are obtained. Afterwards, the vibration amplitude and frequency of the velocity time-history curve decrease. After 30MS, the oscillation amplitude of the velocity time-history curve rises while the oscillation frequency continues to decrease. The maximum values of the vibration velocity during blasting are 1.9976×10^{-2}m/S and 1.5853×10^{-2}m/S respectively, and the minimum values are -2.4304×10^{-2}m/S and -1.7609×10^{-2}m/S respectively. When the stratum is composed of dolomitic limestone, the maximum and minimum values of the vibration velocity are 79.36% and 72.45% respectively of those when the stratum is composed of granite. When the stratum is composed of gravel soil and hard clay, the curve shows a trend of first rising and then falling within 0~60MS while the oscillation amplitude shows a trend of first rising, then falling and then rising again. After 38MS, the oscillation amplitude increases significantly and the maximum and minimum values of vibration velocity in blasting vibration are obtained in about 50MS. The maximum values during blasting are 3.9563×10^{-3}m/S and 1.3088×10^{-3}m/S respectively, which are 19.81% and 6.55% of the maximum values when the stratum is composed of granite. The minimum values of vibration velocity in blasting vibration are -2.0764×10^{-3}m/S and -2.7450×10^{-4}m/S respectively, which are 8.54% and 1.13% respectively of the minimum value of vibration velocity when the stratum is composed of granite.

(3) For the velocity time-history curve in the Y direction, when the stratum is composed of granite and dolomitic limestone, the velocity time-

history curve oscillates violently within 5~15MS and then the oscillation amplitude decreases slowly within 15~29MS. After 29MS, the oscillation amplitude increases again while the oscillation frequency continues to slow down, and the maximum and minimum values of vibration velocity during blasting are obtained. The maximum values of vibration velocity are 1.6895×10^{-2}m/S and 1.8871×10^{-2}m/S respectively, and the minimum values are -2.3701×10^{-2}m/S and -1.9976×10^{-2}m/S respectively. When the stratum is composed of dolomitic limestone, the maximum and minimum values of vibration velocity are -2.3701×10^{-2}m/S and -1.9976×10^{-2}m/S respectively, which are 70% and 37. 04% respectively of the values when the stratum is composed of granite. When the stratum is composed of gravel soil and hard clay, the oscillation frequency of the velocity time-history curve is much lower than that when the stratum is composed of granite and dolomitic limestone. Within 0~60MS, the oscillation amplitude shows a trend of first rising and then falling while the oscillation amplitude first increases, then decreases and then increases again. After 36MS, the oscillation amplitude begins to increase significantly. The maximum vibration velocities of the two strata in blasting vibration are 6.2571×10^{-3}m/S and 2.9155×10^{-3}m/S respectively, which are 37.04% and 17.26% of the maximum vibration velocities when the stratum is composed of granite. The minimum vibration velocities in blasting vibration are -2.3051×10^{-3}m/S and -7.0869×10^{-3}m/S respectively, which are 9.73% and less than one ten-thousandth of the minimum value of vibration velocity when the stratum is composed of granite.

154

(4) For the velocity time-history curve in the Z direction, when the stratum is composed of granite and dolomitic limestone, the oscillation amplitude of the velocity time-history curve increases sharply after 10MS and then the maximum and minimum values of the vibration velocity in the whole blasting vibration are obtained. After 41MS and 32MS respectively, as the oscillation amplitude of the velocity time-history curve decreases, the oscillation frequency decreases significantly. During blasting, the maximum values of the vibration velocity are 7.4537×10^{-4}m/S and 3.4396×10^{-4}m/S respectively and minimum values are -8.0648×10^{-4}m/S and -3.5034×10^{-4}m/S respectively. When the stratum is composed of dolomitic limestone, the maximum and minimum values were 46.15% and 43.44% respectively of those when the stratum is composed of granite. When the stratum is composed of gravel soil and hard clay, the oscillation frequency of the velocity time-history curve is much lower than that when the stratum is composed of granite and dolomitic limestone. Within 0~60MS, the oscillation frequency shows a trend of first rising and then falling while the oscillation amplitude shows a trend of first increasing, then decreasing and then increasing again. The maximum values of the two vibration velocities in blasting vibration are 3.1184×10^{-5}m/S and 8.0401×10^{-6}m/S respectively, which are 4.18% and 2.34% respectively of those when the stratum is composed of granite. The minimum values of the vibration velocities in blasting vibration are -2.2350×10^{-5}m/S and -1.3827×10^{-6}m/S respectively, which are 2.77% and 0.39% of the minimum values of vibration velocity when the stratum is composed of

granite.

(5) When the stratum is composed of granite and weathered dolomitic ash, velocity time-history curve of measuring point #5 in the Z direction shows multiple peaks and troughs. An analysis reveals that the seismic waves transmitted to the measuring point is less affected by the transmission and reflection effects of seismic waves because the measuring point is a certain distance away from the stratum. The velocity time-history curve of the measuring point inherits the characteristics of the velocity time-history curve of measuring point #4 in the Z direction.

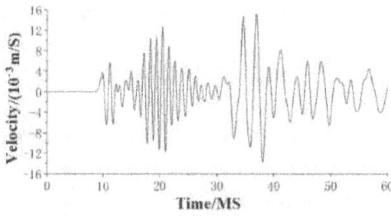

(a) granite interlayer

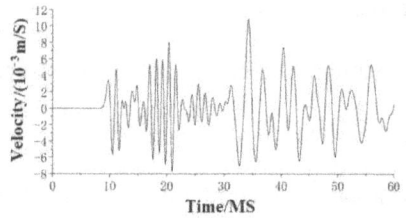

(b) limestone interlayer

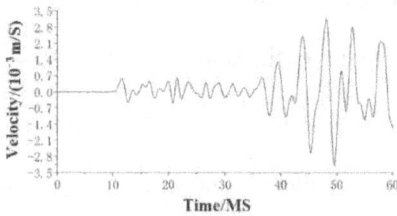

(c) crushed stone interlayer

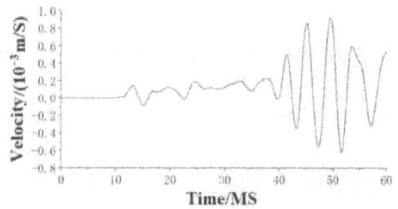

(d) hard clay interlayer

Fig.5.17 Velocity time-history curves in X direction at #6

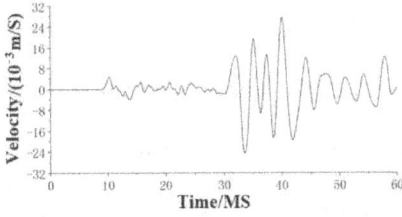

(a) granite interlayer   (b) limestone interlayer

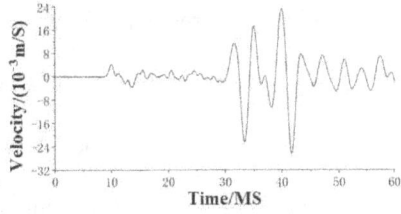

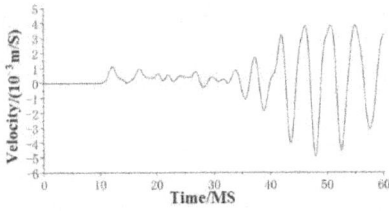

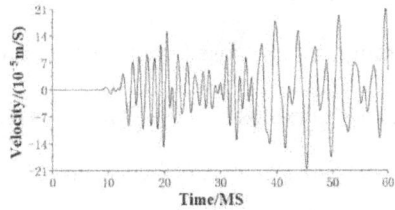

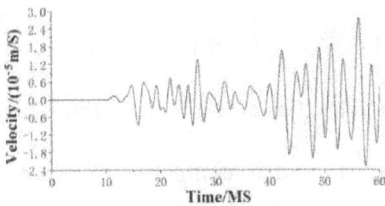

(c) crushed stone interlayer   (d) hard clay interlayer

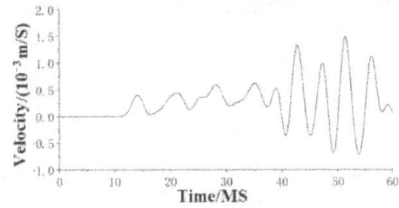

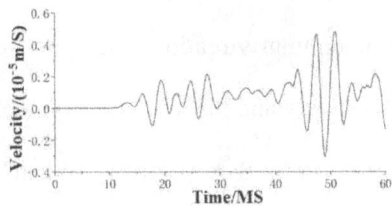

Fig.5.18 Velocity time-history curves in Y direction at #6

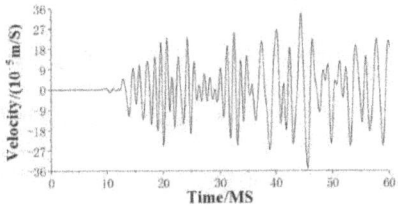

(a) granite interlayer   (b) limestone interlayer

(c) crushed stone interlayer   (d) hard clay interlayer

Fig.5.19 Velocity time-history curves in Z direction at #6

Based on Figures 5.17~5.19, an analysis is conducted on the velocity time-history curves in the X, Y and Z directions at the position with a proportional distance of 15 m/kg$^{1/3}$, yielding the following results:

(1) Under the action of blasting, the hard stratum has little effect on the vibration waveform of measuring point #6 and the effect is mainly reflected in the oscillation amplitude difference of the velocity time-history curve and the timing of the maximum and minimum values of the vibration velocity after 30MS. Bute the weak stratum greatly changes the vibration waveform of the velocity time-history curve of measuring point #6. Compared with the vibration waveform of the hard stratum, the vibration frequency of the seismic waves after passing through the weak stratum is greatly reduced.

(2) For the velocity time-history curve in the X direction, when the stratum is composed of granite and dolomitic limestone, the velocity time-history curve oscillates violently within 11~15MS. Afterwards, the vibration amplitude and frequency of the velocity time-history curve decrease. After 30MS, the oscillation amplitude of the velocity time-history curve increases rapidly while the vibration frequency continues to decrease. During blasting, the maximum vibration velocities are 1.5232×10^{-2}m/S and 1.0881×10^{-2}m/S respectively and the minimum vibration velocities are -1.3713×10^{-2}m/S and -7.6957×10^{-3}m/S respectively. When the stratum is composed of dolomitic limestone, the maximum and minimum values of vibration velocity of are 71.44 % and 12.56% of the values when the stratum is composed of granite. When the stratum is composed of gravel soil and hard clay, the oscillation

amplitude of the velocity time-history curve is small in 8~35MS and it increases significantly after 37MS. The maximum and minimum values of the vibration velocity are obtained in about 50MS, which are 3.1502×10^{-3}m/S and 9.2458×10^{-4}m/S respectively, which are 20.68% and 6.07% of the maximum values of the vibration velocity when the stratum is composed of granite. The minimum values of the two vibration velocities are -3.2030×10^{-3}m/S and -6.2332×10^{-4}m/S respectively, which are 23.36% and 4.55% of the minimum values of the vibration velocity when the stratum is composed of granite.

(3) For the velocity time-history curve in the Y direction, when the stratum is composed of granite and dolomitic limestone, the oscillation amplitude of the velocity time-history curve is small in 8~30MS and increases rapidly after 30MS. The maximum and minimum values of vibration velocity during blasting are obtained in 33~43MS and the oscillation amplitude begins to decay after 45MS. The maximum values of vibration velocity during blasting are 2.7671×10^{-2}m/S and 2.3397×10^{-2}m/S and the minimum values are -2.4413×10^{-2}m/S and -2.6440×10^{-2}m/S respectively. When the stratum is composed of dolomitic limestone, the maximum and minimum values of vibration velocity are -2.4413×10^{-2}m/S and -2.6440×10^{-2}m/S respectively, which are 84.55% and 108.30% respectively of the values when the stratum is composed of granite. When the stratum is composed of gravel soil and hard clay, the oscillation amplitude of the velocity time-history curve is small in 8~38MS and the oscillation amplitude increases rapidly after 38MS. The maximum and minimum values of vibration velocity during blasting are

obtained in 45~55MS. The maximum values of vibration velocity during blasting are 3.8469×10^{-3}m/S and 1.4869×10^{-3}m/S respectively, which are 13.90% and 5.37% respectively of the values when the layer is composed of granite. The minimum values of vibration velocity during blasting are - 4.9263×10^{-3}m/S and -7.1516×10^{-4}m/S respectively, which are 20.18% and 2.93% respectively of the values when the stratum is composed of granite.

(4) For the velocity time-history curve in the Z direction, when the stratum is composed of granite and dolomitic limestone, the oscillation amplitude of the velocity time-history curve increases sharply after 12MS, decreases after 27MS, increases again after 30MS, while the oscillation frequency decreases after 37MS. During the burst period, the maximum values of vibration velocity are 3.3890×10^{-4}m/S and 2.1292×10^{-4}m/S respectively and the minimum values are -3.5135×10^{-4}m/S and -2.0525×10^{-4}m/S respectively. When the stratum is composed of dolomitic limestone, the maximum and minimum values of vibration velocity are 62.83% and 58.42% respectively of the values when the stratum is composed of granite. When the stratum is composed of gravel soil and hard clay, the oscillation frequency of the velocity time-history curve is much lower than that when the stratum is composed of granite and dolomitic limestone. Within 0~60MS, the oscillation amplitude first rises, then decreases, and then rises again. It decreases within 30~40MS and then increases rapidly. After 45MS, the maximum and minimum values of the vibration velocity in the whole blasting vibration are obtained. The maximum values of the vibration velocity are 2.7643×10^{-5}m/S

and 4.8240×10^{-6}m/S respectively, which are 8.16% and 6.44% respectively of the values when the stratum is composed of granite. The minimum values of the vibration velocity are -2.2641×10^{-5}m/S and -3.0783×10^{-6}m/S, which is 2.27% and 1.50% respectively of the values when the stratum is composed of granite.

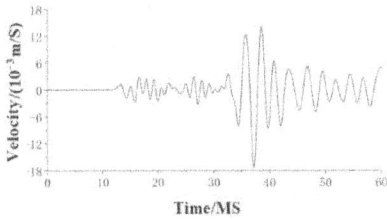

(a) granite interlayer   (b) limestone interlayer

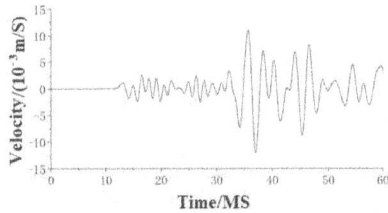

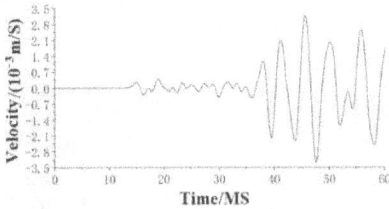

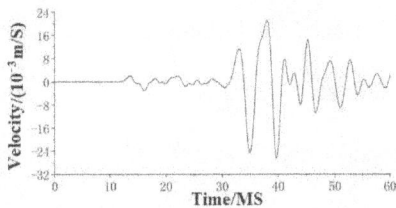

(c) crushed stone interlayer   (d) hard clay interlayer

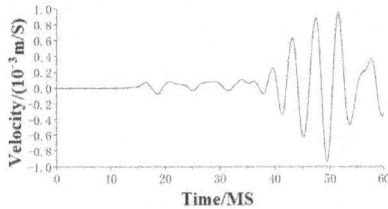

Fig.5.20 Velocity time-history curves in X direction at #7

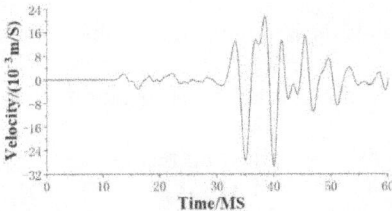

(a) granite interlayer   (b) limestone interlayer

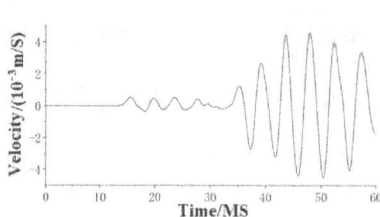

(c) crushed stone interlayer   (d) hard clay interlayer

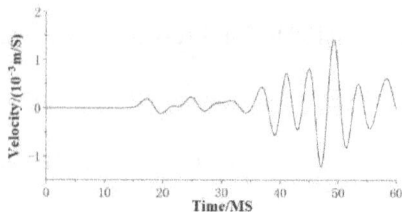

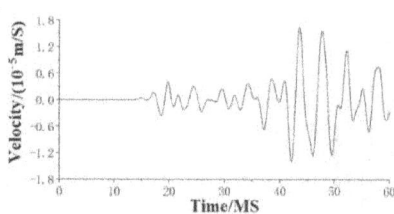

Fig.5.21 Velocity time-history curves in Y direction at #7

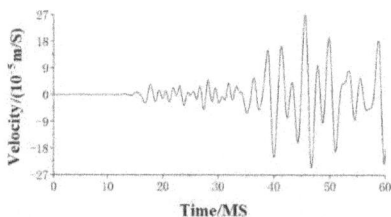

(a) granite interlayer   (b) limestone interlayer

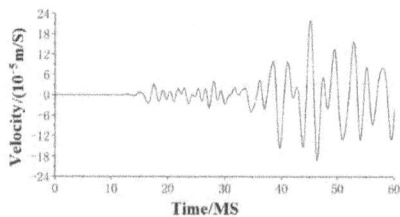

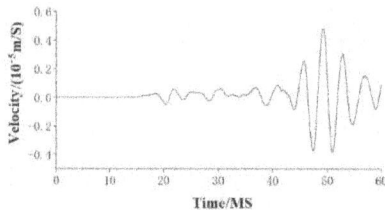

(c) crushed stone interlayer   (d) hard clay interlayer

Fig.5.22 Velocity time-history curves in Z direction at #7

Based on Figures 5.20-5.22, an analysis is conducted on the velocity time-history curves in the X, Y and Z directions at the position with a proportional distance of 20 m/kg$^{1/3}$, yielding the following results:

(1) Under the action of blasting, both hard and weak strata have little effect on the vibration waveform of measuring point #7, and the effect is

mainly reflected in the timing of the maximum and minimum values of blasting vibration velocity during the whole blasting process.

(2) For the velocity time-history curve in the X direction, when the stratum is composed of granite and dolomitic limestone, the oscillation amplitude of the velocity time-history curve is small within 12~32MS and increases rapidly after 32MS. The maximum and minimum values of the vibration velocity during blasting are obtained in about 37MS, and the oscillation amplitude begins to decay after 47MS. The maximum values of the vibration velocity during blasting are 1.4125×10^{-2}m/S and 1.1053×10^{-2}m/S respectively and the minimum values are -1.7194×10^{-2}m/S and -1.2008×10^{-2}m/S respectively. When the stratum is composed of dolomitic limestone, the maximum and minimum values of the vibration velocity are 78.25% and 69.84% of the values when the stratum is composed of granite. When the stratum is composed of gravel soil and hard clay, the oscillation amplitude of the velocity time-history curve is small within 13~38MS and the oscillation amplitude increases rapidly after 38MS. The maximum and minimum values of vibration velocity during blasting are obtained around 49MS. The maximum values of vibration velocity during blasting are 3.1687×10^{-3}m/S and 9.6003×10^{-4}m/S respectively, which are 22.43% and 6.80% of the values when the stratum is granite. The minimum values of vibration velocity during blasting are -3.2880×10^{-3}m/S and -9.3083×10^{-4}m/S respectively, which are 19.12% and 5.41% of the values when the stratum is composed of granite.

(3) For the velocity time-history curve in the Y direction, when the

163

stratum is composed of granite and dolomitic limestone, the oscillation amplitude of the velocity time-history curve is small within 13~38MS and rapidly increases after 32MS. The maximum and minimum values of vibration velocity during blasting are obtained in about 40MS, and the oscillation amplitude begins to decay after 50MS. The maximum values of vibration velocity during blasting are 2.1544×10^{-2}m/S and 2.1043×10^{-2}m/S respectively, and the minimum values are -2.9431×10^{-2}m/S and -2.6631×10^{-2}m/S respectively. When the stratum is composed of dolomitic limestone, the maximum and minimum values of vibration velocity are 97.67% and 90. 49% respectively of the values when the stratum is composed of granite. When the stratum is composed of gravel soil and hard clay, the oscillation amplitude of the velocity time-history curve is small within 13~36MS and it increases rapidly after 36MS. The maximum and minimum values of vibration velocity during blasting are obtained within 43~53MS. The maximum values of vibration velocity during blasting are 4.5923×10^{-3}m/S and 1.4266×10^{-3}m/S respectively, which are 21.32% and 6.62% of the values when the stratum is composed of granite. The minimum values of vibration velocity during blasting are -4.5056×10^{-3}m/S and -1.2153×10^{-3}m/S respectively, which are 15.31% and 4.13% of the values when the stratum is composed of granite.

(4) For the velocity time-history curve in the Z direction, when the stratum is composed of granite and dolomitic limestone, the oscillation amplitude of the velocity time-history curve is small in 12~32MS and rapidly increases after 36MS. The maximum and minimum values of vibration

164

velocity during blasting are obtained in about 45MS. The maximum values of vibration velocity during blasting are 2.7438×10^{-4}m/S and 2.2116×10^{-4}m/S respectively and the minimum values are -2.4326×10^{-4}m/S and -1.9219×10^{-4}m/S respectively. When the stratum is composed of dolomitic limestone, the maximum and minimum values of vibration velocity are 80.61% and 79.00% respectively of the values when the stratum is composed of granite. When the stratum is composed of gravel soil and hard clay, the oscillation amplitude of the velocity time-history curve is small within 13~40MS, and the oscillation amplitude increases rapidly after 40MS. The maximum and minimum values of vibration velocity during blasting are obtained within 42~52MS. The maximum values of vibration velocity during blasting are 1.6635×10^{-5}m/S and 4.8057×10^{-6}m/S respectively, which are 6.06% and 2.17% respectively of values when the stratum is composed of granite. The minimum values of vibration velocity during blasting are -1.3904×10^{-5}m/S and -3.8555×10^{-6}m/S respectively, which are 5.72% and 2.01% respectively of the values when the stratum is granite.

5.2 Comparative Analysis of Blasting Vibration

Based on the above numerical simulation calculation, the time-history changes of blasting vibration velocities in the X, Y and Z directions of measuring points #1~#7 under different stratum inclusions are obtained. To further explore the relationship between the maximum and minimum values of blasting vibration velocity in each vibration direction and the proportional

distance, the following analysis is carried out:

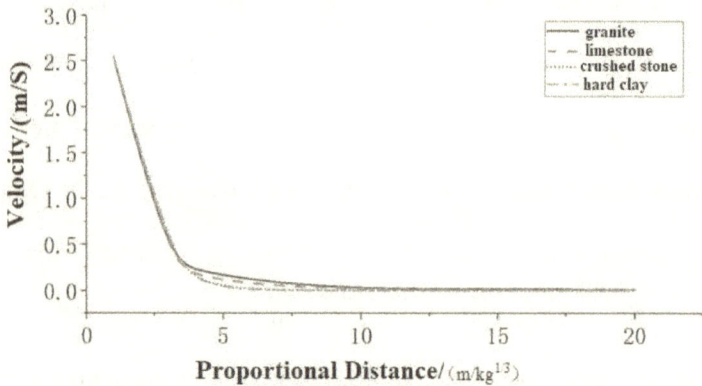

(a) Maximum blasting vibration velocity

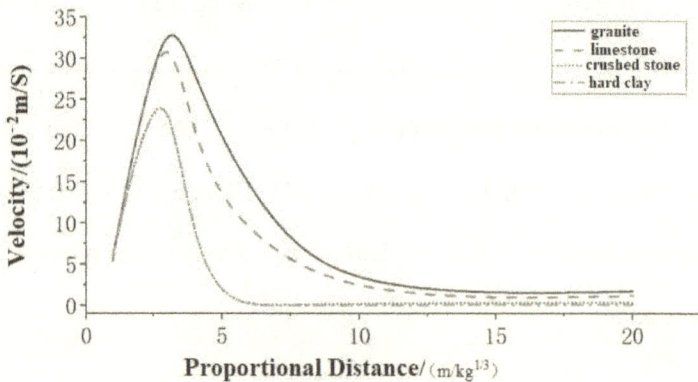

(b) Minimum blasting vibration velocity

Fig.5.23 Blasting vibration in X direction with proportional distance

It can be seen from Figure 5.23 that the maximum value of blasting

vibration velocity in the X direction at each measuring point with different

stratum inclusions is basically the same, and it is affected only within the proportional distance of 3~8m/kg$^{1/3}$. Therefore, it can be concluded that when 25kg explosive is used, only the near strata significantly affect the maximum value of blasting vibration velocity in the X direction. Since 25kg explosive is used in this simulation, the blasting vibration in the X direction in the near area is mainly directed away from the direction of the explosion source, resulting in a small minimum vibration velocity in the near area of blasting. By mainly referring to the minimum blasting vibration velocity in the X direction after 3 m/kg$^{1/3}$, it can be found that when the stratum is composed of dolomitic limestone, the minimum blasting vibration velocity in the X direction is significantly lower than that in other strata, while the minimum blasting vibration velocity curve in the X direction when the stratum is composed of gravel soil and hard clay almost overlap and lie below the curve when the stratum is composed of dolomitic limestone. All the four curves eventually approach zero. Therefore, it is concluded that when 25kg explosive is used, both hard and weak strata reduce the minimum value of blasting vibration velocity in the X direction, but the weak stratum have a pronounced reducing effect on the minimum value of blasting vibration velocity compared to the hard stratum.

As shown in Figure 5.24, the maximum value curve of blasting vibration velocity in the Y direction of each measuring point with different stratum inclusions is very close and it is affected only within the proportional distance of 3~8m/kg$^{1/3}$.

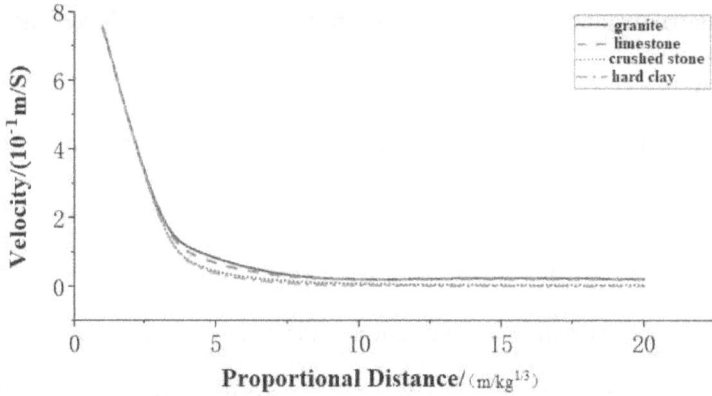

(a) Maximum blasting vibration velocity

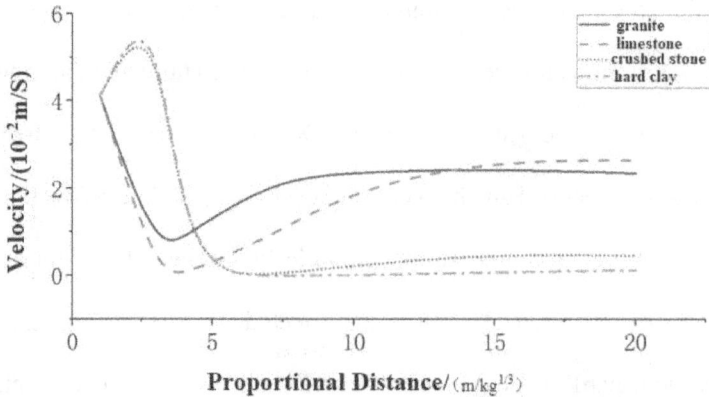

(b) Minimum blasting vibration velocity

Fig.5.24 Blasting vibration in X direction with proportional distance

Therefore, it can be concluded that when 25kg explosive is used, the strata affect the maximum value of blasting vibration velocity in the Y direction nearby and the weak stratum has the maximum value of blasting vibration velocity in the Y direction. Because 25kg explosive is used in this simulation, the blasting vibration in the Y direction in the near area is mainly directed away from the direction of the blasting source, resulting in a small minimum vibration in the near area of blasting. This makes the minimum value curve of the blasting velocity in the Y direction not a decreasing curve. It can be seen from the figures that all the four curves have a first rising and then falling process. When the stratum is composed of gravel soil or hard clay, this trend occurs before crossing the stratum. When the stratum is composed of gravel soil or hard clay, this trend occurs after crossing the stratum. In addition, when the stratum is composed of gravel soil or hard clay, the minimum value of blasting vibration velocity in the Y direction decreases much faster than when the layer is composed of granite pyro dolomitic limestone.

As shown in figure 5.25 that the maximum value curve of blasting vibration velocity in Z direction at each measuring point with different stratum inclusions is basically the same, and it is affected only within the proportional distance of $3{\sim}8m/kg^{1/3}$.

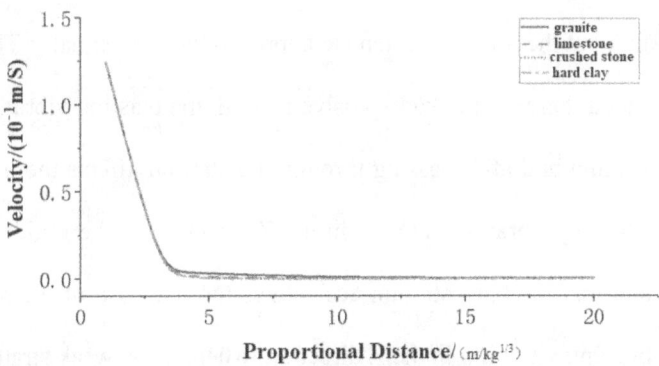

(a) Maximum blasting vibration velocity

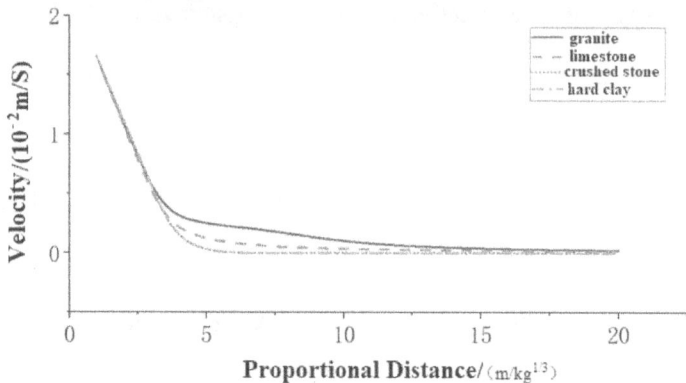

(b) Minimum blasting vibration velocity

Fig.5.25 Blasting vibration in X direction with proportional distance

Therefore, it can be concluded that when 25kg explosive is used, the maximum value of blasting vibration velocity in the Z direction is significantly affected only near the strata. It can be found from the figure that the minimum value curve of blasting vibration velocity in the Z direction at each measuring point with different stratum inclusions is roughly the same, and the decreasing speed is different only when the proportional distance is $3\sim15\text{m/kg}^{1/3}$, but the four curves tend to approach 0m/S eventually. Therefore, it is concluded that when 25kg explosive is used, the blasting vibration wave near the stratum and after passing through the stratum affects the maximum value of blasting vibration velocity in the Z direction in the middle and far regions, and that the hard stratum accelerates the decrease of the minimum value of blasting vibration in the Z direction, whereas the weak stratum has a greater impact on the decreasing speed of the minimum value of blasting vibration in the Z direction.

5.3 Simulation of Blasting Displacement

The blasting vibration displacement time-history curves of measuring points #1~#7 with the proportional distances of $1m/kg^{1/3}$, $3m/kg^{1/3}$, $4m/kg^{1/3}$, $7m/kg^{1/3}$, $10m/kg^{1/3}$, $15m/kg^{1/3}$ and $20m/kg^{1/3}$ along the horizontal radial direction of the concentrated charge are extracted and analyzed to study the influence of different stratum inclusions on the propagation law of blasting seismic waves in rock masses. Measuring point #1 displacement time-history curves are not to be analyzed since they are highly overlapping.

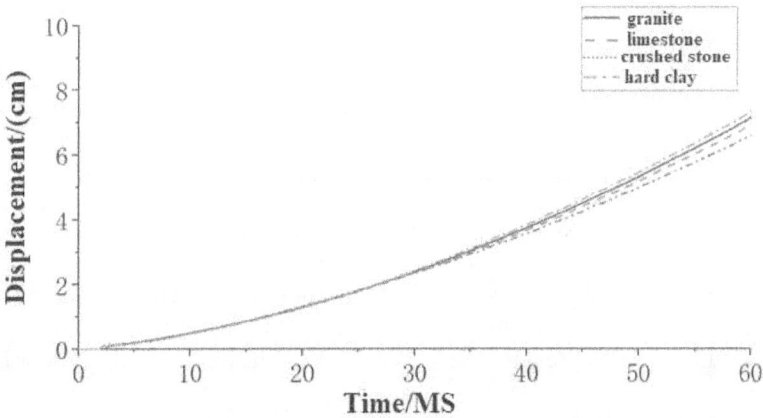

Fig.5.26 Displacement time-history curves at #2

According to Figure 5.26, an analysis is conducted on the displacement time-history curve of measuring point #2 with a proportional distance of $3m/kg^{1/3}$. The measuring point is located between the explosion source and the stratum and is relatively close to the stratum. The displacement time-history curve of the measuring point exhibits an increasing trend with a gradually rising slope. The maximum value of the four curves is obtained at 60MS within the simulation range of 0~60MS. Within 0~30MS, the displacement time-history curves of the four different strata almost overlapping and then the four displacement time-history curves begin to diverge significantly. When the stratum is composed of dolomitic limestone or gravelly soil, the displacement time-history curve of measuring point #2 at 60MS is obviously smaller than that when the stratum is composed of granite, which is 96.90% and 92.18% of that when the stratum is granite, respectively. When the stratum is composed of hard clay, the displacement time-history curve of measuring point #2 at 60MS is significantly larger than that when the stratum is composed of granite, which is 102.57% of that when the stratum is composed of granite. From 2.3.3, it can be seen that the elastic modulus of dolomitic limestone used in this study is 30GPa; the elastic modulus of granite is 500MPa; the elastic modulus of gravelly soil is 140MPa, and the elastic modulus of hard clay is 12MPa. It can be inferred that when the strength of the stratum is much greater than that of the rock masses on both sides or is an order of magnitude with the elastic modulus of the rock masses on both sides, the final displacement of the measuring point located between the explosion source and the stratum and close to the stratum is smaller than that without stratum. This is because the interface makes the blasting seismic waves reflect and transmit. The reflected seismic waves offset a small part of the subsequent seismic waves transmitted here, making the final displacement of measuring

172

point #2. When the strength of the stratum is far less than the rock mass on both sides, the final displacement of the measuring point located between the explosion source and the stratum and close to the stratum is larger than that in the absence of stratum. This is because the weak stratum cannot resist the impact of blasting seismic waves, which makes the displacement of the weak stratum larger than that of the rock mass on both sides, and increases the final displacement under the action of rock mass extrusion.

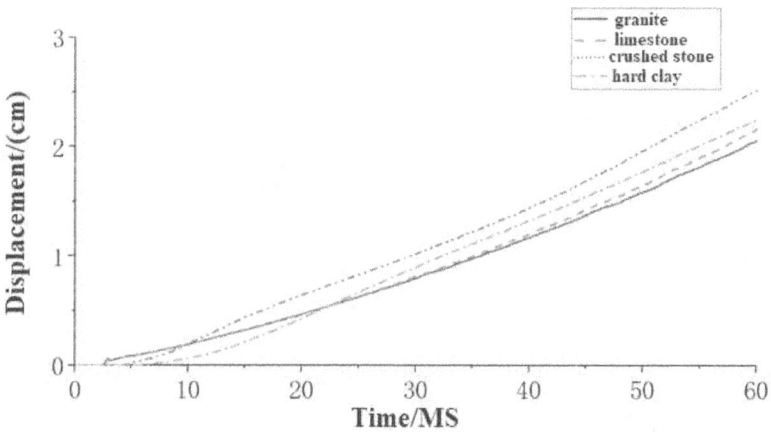

Fig.5.27 Displacement time-history curves at #3

From Figure 5.27, an analysis is conducted on the displacement time-history curve at measuring point #3 with a proportional distance of $4m/kg^{1/3}$. The measuring point is located on the right side of the stratum. At this time, the seismic waves have passed through the stratum and the displacement time-history curve of the measuring point is increasing, and the maximum value is obtained in 60MS within the simulation range of 0~60MS. When the stratum is composed of granite or dolomitic limestone, the two displacement time-history curves keep increasing and are almost overlapping in 0~30MS. When the stratum is composed of dolomitic limestone, the displacement time-history curve increases rapidly at 30MS, making the displacement of the measuring point at 60MS is greater than that when the stratum is composed of granite, which is 105.04% of that when the stratum is composed of granite. When the stratum is composed of gravel soil or hard clay, the displacement time at the measuring point is significantly later than that when the stratum is compose of granite, but the displacement growth rate is significantly faster than that when the stratum is granite. The displacement of the measuring point at 60MS is greater than that when the stratum is composed of granite, which is 122.54% and 109.43% of that when the stratum is granite, respectively. It can be inferred that the displacement of the measuring point near the right side of the stratum can be significantly inhibited in the early stage of the explosion in the weak stratum, but the displacement growth rate of the measuring point is faster after the displacement of the measuring point occurs. In addition, whether the strength of the stratum is hard or weak, the final displacement of the measuring point near the right side of the stratum is greater than that when there is no stratum.

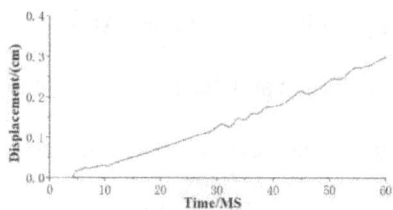

(a) granite interlayer   (b) limestone interlayer

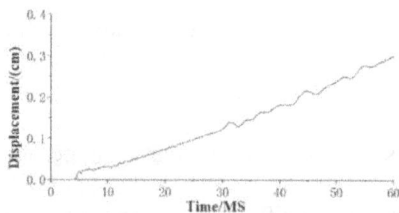

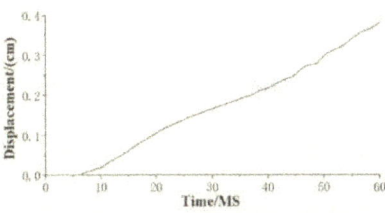

(c) crushed stone interlayer   (d) hard clay interlayer

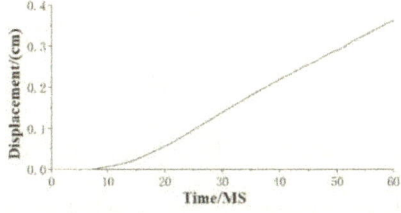

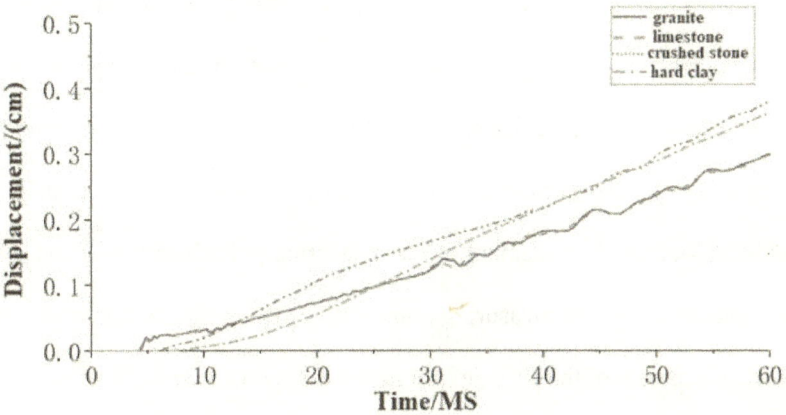

(e) #4 Comparison of displacement

Fig.5.28 Displacement-history curves at position #4

According to Figure 5.28, an analysis is conducted on the displacement time-history curve of measuring point #4 with a proportional distance of $7m/kg^{1/3}$. The measuring point is located on the right side of the stratum and the seismic waves have passed through the stratum and propagated for a certain distance. When the stratum is composed of gravel soil or hard clay, the displacement time-history curve at measuring point #4 shows an increasing trend, and the displacement increase is relatively stable. The maximum value is obtained at 60MS in the simulation range of 0~60MS, and the final

175

displacement is 127.57% and 120.83% respectively when the stratum is composed of granite. When the stratum is composed of dolomitic limestone, the displacement time-history curve at measuring point #4 is close to that when the stratum is composed of granite, and both of them show an increasing trend within 0~30MS. After 30MS, the displacement first increases rapidly and then decreases, but the maximum displacement is obtained at 60MS and the final displacement is 99.60% of that when the stratum is composed of granite. It can be inferred that the weak stratum can significantly restrain the displacement of the particle near measuring point #4 at the initial stage of explosion, but the displacement of the measuring point increases faster after the displacement of the measuring point occurs. The weak stratum makes the final displacement of the particle near measuring point #4 more than 1.2 times when the stratum is composed of granite. The hard stratum has little influence on the final displacement of the particle near measuring point #4 and its displacement time-history curve.

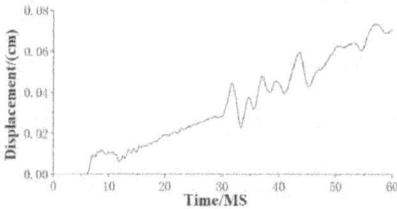

(a) granite interlayer　　　　　　(b) limestone interlayer

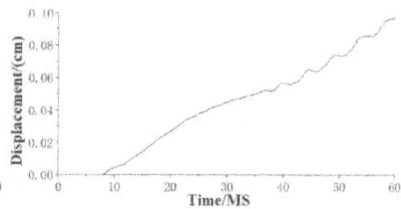

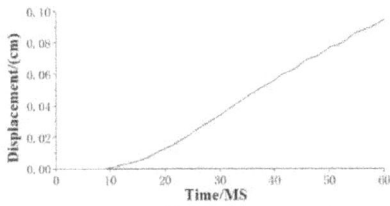

(c) crushed stone interlayer   (d) hard clay interlayer

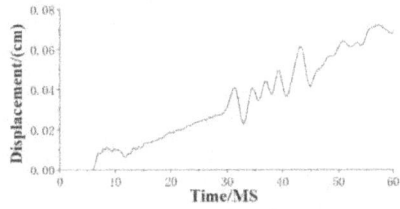

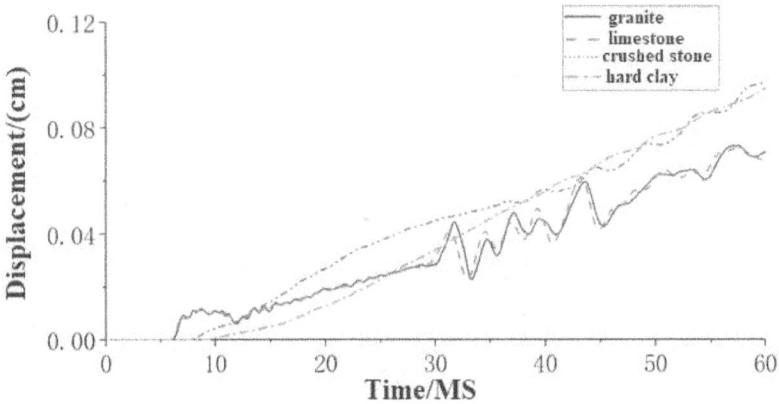

(e) #5 Comparison of displacement

Fig.5.29 Displacement time-history curves at #5

Based on Figure 5.29, an analysis is conducted on the displacement time-history curve of measuring point #5 with a proportional distance of $1010m/kg^{1/3}$. The measuring point is located on the right side of the stratum and the seismic waves have passed through the stratum and propagated for a certain distance. When the stratum is composed of gravel soil or hard clay, the displacement time-history curve shows an increasing trend, and the displacement increase is relatively stable in 0~35MS. After 35MS, the displacement time-history curve fluctuates when the stratum is composed of gravel soil, but the maximum value is obtained in 60MS in the simulation range of 0~60MS, and the final displacement is 132.63% and 129.91% respectively of that when the stratum is composed of granite. When the stratum is composed of dolomitic limestone, the displacement time-history curve at measuring point #5 is highly similar to the curve when the stratum is composed of granite, and both fluctuate and rise within 6~16MS. The displacement increase becomes gentle after 16MS and the displacement fluctuation amplitude increases suddenly after 30MS but begins to decrease after 45MS. The maximum value is obtained in about 57MS within the simulation time range of 0~60MS. The maximum displacement is 98.82% of that when the stratum is composed of granite. It can be inferred that the weak stratum can obviously restrain the displacement of the particle near measuring point #5 at the initial stage of explosion, but the displacement growth rate will be faster after the displacement of the measuring point. The final displacement of the particle near measuring point #5 is about 1.3 times of the final displacement when the stratum is composed of granite. The hard stratum has little effect on the final displacement of the particle near measuring point #5 and its displacement time-history curve.

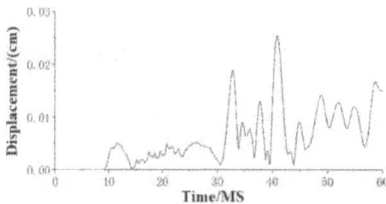

(a) granite interlayer   (b) limestone interlayer

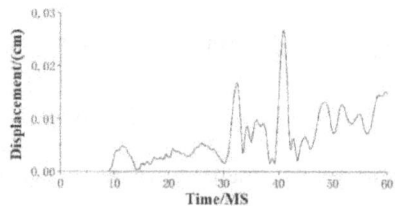

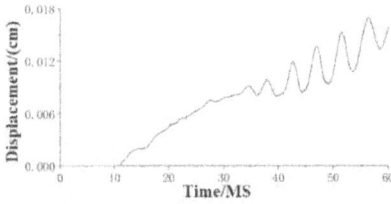

(c) crushed stone interlayer　　　　　(d) hard clay interlayer

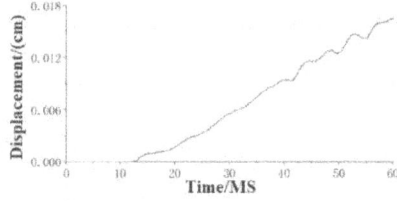

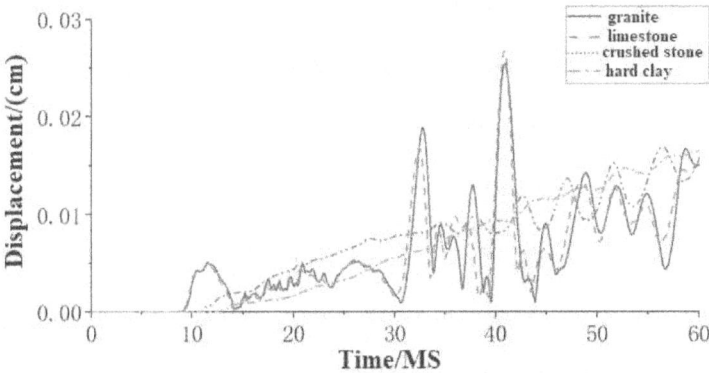

(e) #6 Comparison of displacement

Fig.5.30 Displacement time-history curves at #6

According to Figure 5.30, an analysis is conducted on the displacement time-history curve of measuring point #6 with a proportional distance of $1510m/kg^{1/3}$. The measuring point is far away from the stratum and the seismic waves have passed a long distance. When the stratum is composed of hard clay, the displacement time-history curve shows an increasing trend. After 40MS, the displacement time-history curve fluctuates to a certain extent. Within the simulation range of 0~60MS, the maximum value is obtained at 60MS, and the maximum displacement is 65.55% of that when the stratum is composed of granite. When the stratum is composed of gravel soil, the displacement time-history curve rises steadily in 10~35MS and fluctuates greatly after 35MS. The maximum value is obtained in about 57MS in the simulation time range of 0~60MS, and the maximum displacement is 66.42% of that when the stratum is composed of granite. The displacement time-history curve when the stratum is composed of dolomitic limestone is roughly similar to that when the stratum is composed of granite. The displacement increases sharply and then decreases sharply within the range of 9~15MS. The displacement fluctuation increases after 15MS and the amplitude increases suddenly after 30MS and then decreases after 45MS. The maximum displacement is 104.77% of that when the stratum is composed of granite within the simulation range of 0~60MS. It can be concluded that the weak stratum can obviously restrain the displacement of the particle near measuring point #6 at the initial stage of explosion, and the displacement of the measuring point increases steadily after the displacement of the measuring point occurs. The hard stratum has little effect on the time-history curve of particle displacement near measuring point #6.

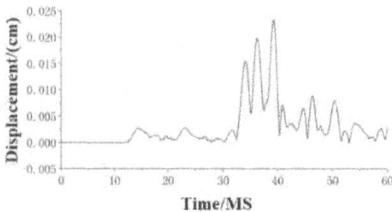

(a) granite interlayer   (b) limestone interlayer

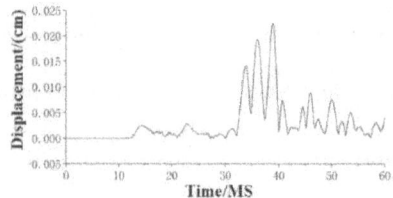

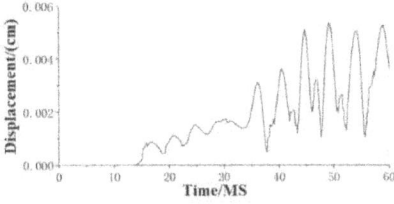

(c) crushed stone interlayer   (d) hard clay interlayer

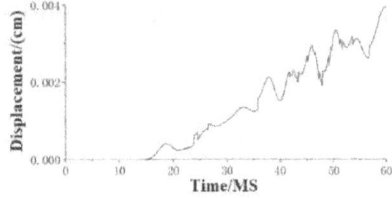

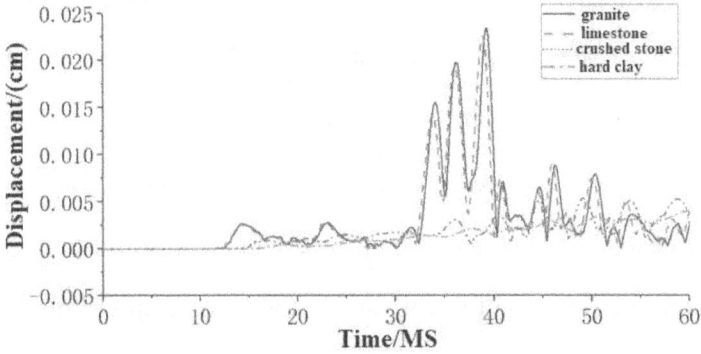

(e) #7 Comparison of displacement

Fig.5.31 Displacement time-history curves at #7

According to Figure 5.31, an analysis is conducted on the displacement time-history curve of measuring point #7 with a proportional distance of 20 $m/kg^{1/3}$. The measuring point is far away from the stratum and the seismic waves have propagated for a considerable distance. When the stratum is composed of hard clay, the displacement time-history curve shows an increasing trend. After 35MS, the displacement time-history curve fluctuates to a certain extent and the maximum displacement is 16.91% of that when the stratum is composed of granite in the simulation time range of 0~60MS. When the stratum is composed of gravel soil, the displacement time-history curve

181

rises steadily in 15~35MS and fluctuates greatly after 35MS. The maximum value is obtained in about 59MS in the simulation range of 0~60MS, and the maximum displacement is 22.94% of that when the stratum is composed of granite. The displacement time-history curve when the stratum is composed of dolomitic limestone is roughly similar to that when the stratum is composed of granite. Both of them fluctuate in a small positive range within 12~25MS and produce a negative displacement after 25MS. The displacement time-history curve begins to fluctuate violently after 32MS and its fluctuation amplitude begins to decrease after 40MS. In this process, the maximum displacement in the whole simulation process is generated, and the maximum displacement is 96.36% of that when the stratum is composed of granite. It can be concluded that the weak stratum can obviously restrain the displacement of the particle near measuring point #7 at the initial stage of explosion and the displacement of the measuring point increases steadily at the beginning after the displacement of the measuring point occurs. The hard stratum also has little effect on the displacement time-history curve of the particle near measuring point #7. Different from the displacement time-history curve of the measuring point with a proportional distance less than 10 m/kg$^{1/3}$, the maximum displacement of the measuring point in the simulation range of 0~60MS is not obtained in about 60MS. The maximum displacement is not the final displacement obtained by the measuring point and the final displacement is likely to be less than the maximum displacement.

Through comprehensive analysis of Figures 5.26~5.31, it can be found that within the simulation time range of 0~60MS, for the measuring points located in the middle of the explosion source and the stratum and close to the stratum, if the magnitude of the strength of the rock mass on both sides is greater than or equal to the magnitude of the strength of the stratum, the particle displacement decreases, but it increases when the strength of the stratum is very small. When the blasting seismic waves cross the stratum and the proportional distance is within the range of 3~10 m/kg$^{1/3}$, the displacement of particles can be significantly inhibited in the weak stratum at the initial stage of explosion, but the growth rate of displacement becomes faster after displacement, and the final displacement increases. The hard stratum has little effect on the final displacement of nearby particles and the displacement time-history curve. The maximum displacement is generally less than that when the stratum is composed of granite. When the blasting seismic waves cross the stratum and the proportional distance exceeds 15 m/kg$^{1/3}$, the displacement of nearby particles can be significantly inhibited in the weak stratum at the initial stage of explosion. After the displacement of the measuring point occurs, the displacement of the measuring point increases steadily, and the maximum displacement generated by the blasting seismic waves is smaller than that when the stratum is composed of granite. With the increase of the proportional distance, the maximum displacement generated is smaller than that when the stratum is composed of granite. The hard stratum has little effect on the displacement time-history curve of nearby particles.

5.4 Comparative Analysis of Strata Displacement

Measuring points #8~#10 are selected on the interface between the right side of the model stratum and the rock mass, which are 0m, 2m and 4m away from the ground, as shown in Figure 5.32. The time-history curves of blasting displacement at the three measuring points were extracted and analyzed to study the effect of different strata inclusions on the propagation law of blasting seismic waves in rock masses.

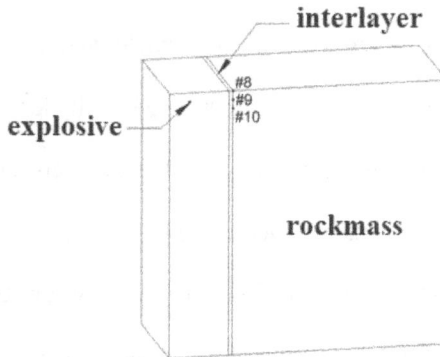

Fig.5.32 Location map of rock mass measuring points

Based on Figure 5.33, an analysis is made on the displacement time-history curve at measuring point #8, which is located at the interface of stratum and rock mass on the free surface. The displacement time-history curve of the measuring point increases gradually and the slope increases gradually.

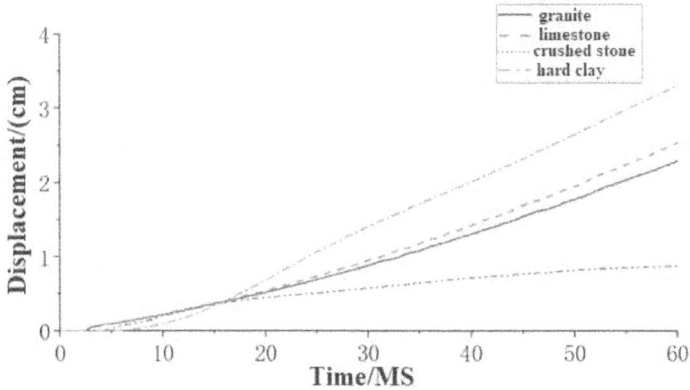

Fig.5.33 Displacement-history curves at #8

Within the simulation range of 0~60MS, the four displacement time-history curves reach the maximum values at 60MS. When the stratum is composed of dolomitic limestone, the displacement time-history curve in the range of 0~20MS almost overlap with that when the stratum is composed of granite. After 20MS, the displacement increase rate is faster than that when the stratum is composed of granite, and the maximum displacement is 110.80% of that when the stratum is composed of granite. When the stratum is composed of gravel soil, the time at which displacement occurs of the measuring point is later than that when the stratum is composed of granite. The increase rate of the displacement of the measuring point is always less than that when the stratum is composed of granite, and the maximum displacement is 38.20% of that when the stratum is composed of granite. When the stratum is composed of hard clay, the displacement of the measuring point is later than that when the stratum is composed of granite, and the displacement increases very fast after displacement. After 17MS, the speed of the displacement time-history curve has been above the displacement time-history curve when the stratum is composed of granite, and the maximum displacement is 144.42% of that when the stratum is composed of granite. The elastic modulus of dolomitic limestone, granite, gravelly soil and hard clay used in this calculation is 30GPa, 500MPa, 40MPa, and 12MPa respectively. It can be inferred that within the simulation time range of 0~60MS, when the magnitude of the strata strength is greater than or less than that of the rock masses on both sides, the displacement of the measuring point increases; when the magnitude of the strata strength is the same as that of the rock masses on both sides, the displacement of the measuring point decreases. The existence of weak strata inhibits the displacement of the measuring point at the initial stage, and the weaker the stratum is, the longer the inhibition time becomes.

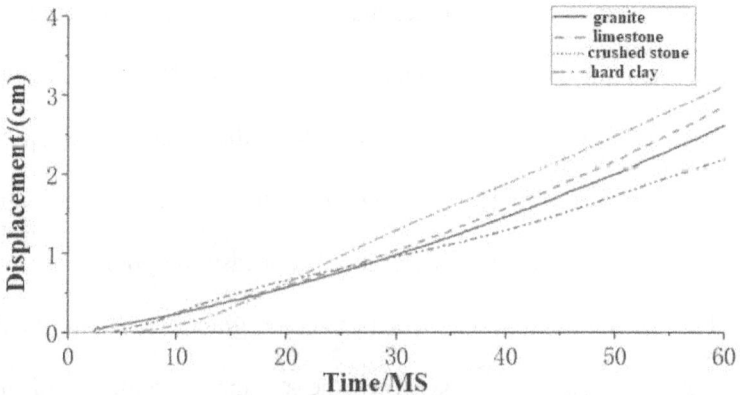

Fig.5.34 Displacement-history curves at #9

Based on Figure 5.34, an analysis is made on the displacement time-

186

history curve of measuring point #9, which is located on the interface of the stratum and rock mass. The measuring point is 2m away from the free face and on the same level with the explosives. The displacement time-history curve of the measuring point shows an increasing trend with a gradually rising slope. Within the simulation range of 0~60MS, the four displacement time-history curves reach the maximum value at 60MS. When the stratum is composed of dolomitic limestone, the displacement time-history curve almost overlap with that when the stratum is composed of granite in the range of 0~20MS. After 20MS, the displacement increase rate is faster than that when the stratum is composed of granite, and the maximum displacement is 109.34% of that when the stratum is composed of granite. When the stratum is composed of gravel soil, the time at which displacement of the measuring point occurs is later than that when the stratum is composed of granite. Within the range of 10~27MS, the displacement time-history curve is above the displacement time-history curve when the stratum is composed of granite. On the contrary, after 27MS, the maximum displacement is 83.63% of that when the stratum is composed of granite. When the stratum is composed of hard clay, the displacement of the measuring point is later but increases more rapidly after the beginning of displacement than that when the stratum is composed of granite. After 17MS, the rate of the displacement time-history curve has been above the displacement time-history curve when the stratum is composed of granite, and the maximum displacement is 119.00% of that when the stratum is composed of granite. The elastic modulus of dolomitic limestone, granite, gravelly soil and hard clay is 30GPa, 500MPa, 140MPa, and 12MPa respectively. It can be inferred that within the simulation time range of 0~60MS, when the magnitude of the strata strength is greater than or less than the rock masses on both sides, the displacement of the measurement point increases, and when the magnitude of the strata strength is the same as the rock masses on both sides, the final displacement of the measuring point decreases. The existence of weak stratum inhibits the displacement of the measuring point at the initial stage, and the weaker the stratum is, the longer the inhibition time becomes.

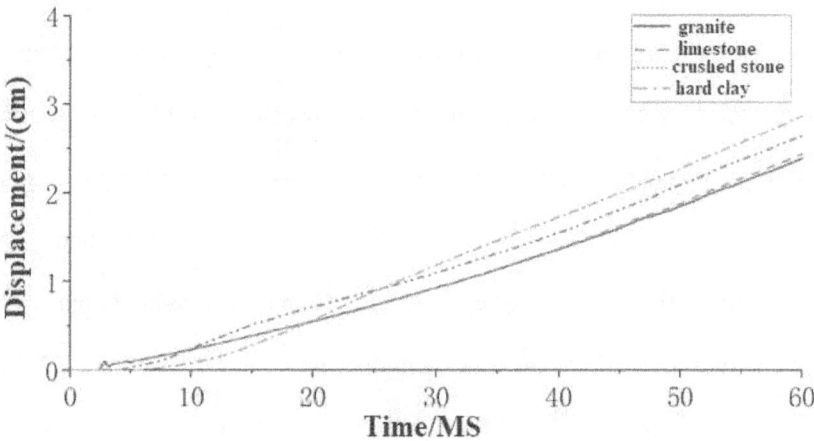

Fig.5.35 Displacement-history curves at #10

Based on Figure 5.35, an analysis is made on the displacement time-history curve of measuring point #10, which is located on the interface of the stratum and rock mass, 4m away from the free face. The displacement time-history curve of the measuring point shows an increasing trend with a gradually rising slope. Within the simulation range of 0~60MS, the four displacement time-history curves reach the maximum value at 60MS. When the stratum is composed of dolomitic limestone, the displacement time-history curve almost overlaps with that when the stratum is composed of granite in the range of 0~20MS. After 45MS, the displacement increase rate is faster than that when the stratum is composed of granite, and the maximum displacement is 102.29% of that when the stratum is composed of granite. When the stratum is composed of gravel soil, the displacement time of the measuring point is later than that when the stratum is composed of granite, but the increase rate of displacement after the beginning is much faster than that when the stratum is composed of granite. The maximum displacement is 110.51% of that when the stratum is composed of granite. When the stratum is composed of hard clay, the displacement of the measuring point is later than that when the stratum is composed of granite, and the increase rate of displacement after the beginning is much faster than that when the stratum is composed of granite. After 19MS, the rate of displacement time-history curve has been above the displacement time-history curve when the stratum is composed of granite, and the maximum displacement is 119.97% of that when the stratum is composed of granite. It can be inferred that within the simulation time range of 0~60MS, whether the stratum is weak or hard, the existence of the strata increases the final displacement at the measuring point. The existence of weak stratum restrains the displacement of the measuring point at the initial stage, and the weaker the stratum is, the longer the time it can be restrained, and the greater the final displacement at the measuring point.

5.5 Engineering Verification

1. An Overview

According to the preliminary analysis of the above numerical simulation, combined with the engineering example, the numerical simulation results are compared with the measured results in the actual project. In this study, an earth rock blasting project in Tangshan is taken as an example. The blasting vibration is monitored on site and the monitoring data are analyzed and compared with the numerical simulation to explore the particle propagation law under the blasting effect.

The project is located in Tangshan City, Hebei Province. The experimental mining area is a part of the northwest wing of Kaiping syncline, and the stratum trend of the mining area is northeast. Under the action of NW horizontal lateral pressure, the stratum becomes steep and partially inverted, and a series of compressional torsional or tensional torsional faults and traction folds are generated. It has been found that there are 17 faults with a certain scale in the integration area, most of which are perpendicular to the trend of the orebody, nearly parallel, with roughly the same spacing, relatively uniform distribution, and horizontal displacement. The orebody is not obviously affected by the faults, and no wide fracture zone and alteration are found [76].

The orebody trend in the mining area is generally north east, and the orebody is mostly layered, 5km long and 200~600m wide. The orebody is buried deeply and the maximum burial depth is below ~100m according to the control of drilling engineering.

The terrain of the mining area is mainly limestone hills. There are few high mountains inside the mining area, with the highest elevation of 283.6m and the lowest elevation of 50.8m. The overall terrain of the mining area is relatively flat, and there are some gullies in the northern part. Because Douhe reservoir is located in the southeast of the mining area, it has a certain effect on the groundwater level in the mining area, but generally speaking, the hydrogeological conditions are simple.

This blasting project is carried out in the mining area as planned. The site picture is shown in Figure 5.36.

Fig.5.36 Site construction photos

The blasting vibration monitoring equipment is used to monitor the vibration characteristics of blasting engineering in Tangshan mining area, and the blasting vibration monitoring data of each measuring point are recorded at the same time. This time, TC-4850 blasting vibration meter is used to monitor the blasting engineering site.

The millisecond delay hole-by-hole blasting method is adopted for the open-pit mine observed in the engineering test, and the specific blasting parameters are as follows.

Blasting layout parameters: $a \times b = (5 \sim 6) \times (4 \sim 5)$m;

Blasting hole diameter: d=310mm;

Emptying and filling length: L_t=6.0~6.5m;

Blast hole depth: L=14.5m; Super depth of blast hole: h=2.5m;

The explosives used for blasting are ANFO and emulsion explosives. Explosive consumption for blasting: q=0.5~1.1kg/m^3.

According to the actual project, six monitoring points are selected to analyze the blasting vibration velocity, and the corresponding proportional distance is calculated according to formula (3.5), as shown in Table 5.1.

Table 5.1 Parameters of selected blasting vibration monitoring points

Position	1	2	3	4	5	6
Distance (m)	101	134	172	219	265	312
Charge mass (kg)	180	180	180	180	180	180
Proportional distance	17.89	23.73	30.46	38.79	46.93	55.26

The effective monitoring values of each measuring point can be obtained from the blasting data recorded by the vibration meter in the field project. The monitoring data of each measuring point are as follows:

Table 5.2 Data of measuring point

Position	1	2	3	4	5	6
Ch.X vibration (cm/S)	1.66	0.38	0.7	1.52	1.11	0.32
Ch.Y vibration (cm/S)	3.23	0.43	1.3	0.56	0.86	0.51
Ch.Z vibration (cm/S)	0.72	0.08	0.39	0.22	0.44	0.16
Ch.X Dominant freq. (Hz)	12.79	12.27	12.47	14.76	6.43	5.35
Ch.Y Dominant freq. (Hz)	12.24	9.77	21.05	22.62	6.05	7.01
Ch.Z Dominant freq. (Hz)	11.83	19.05	10.34	8.79	10.5	5.94

The maximum values of blasting vibration velocity in the horizontal radial, horizontal tangential and vertical directions and the main frequency of blasting vibration in the X direction recorded at each monitoring point in the engineering example are plotted according to the proportional distance. The obtained images are compared with those obtained by plotting the maximum values of blasting vibration velocity in the horizontal radial, horizontal tangential and vertical directions and the main frequency of blasting vibration in the X direction recorded at each monitoring point in the numerical simulation to verify the accuracy of the numerical model.

The maximum value of blasting vibration velocity is compared by proportional distance, and the change of blasting vibration peak value with the increase of proportional distance is obtained, as shown in Figure 5.1.

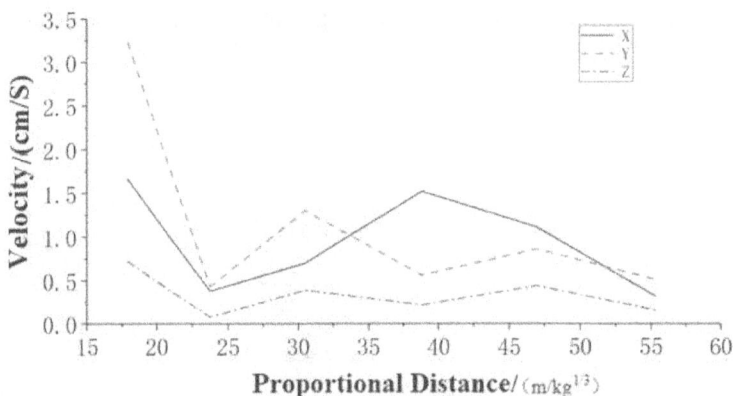

Fig.5.37 Variations of blasting maximum vibration velocity with proportional distance

It can be seen from Figure 5.37 that when the explosive dosage is 180kg, the particle vibration peaks in the X, Y and Z directions gradually attenuate with the increase of the proportional distance as a whole, but the attenuation amplitude in each direction is different and due to geological conditions, the particle vibration peaks of some measuring points increase. When the proportional distance $\overline{R}$ is in the range of 15~25 m/kg$^{1/3}$, the blasting vibration velocity in the Y direction Is the maximum, which is 3.23cm/S, and the blasting vibration velocity in the x direction is the maximum, which is 1.66cm/S. The blasting particle vibration peak in the Z direction is the minimum, which is 0.72cm/S, and the variation amplitude of particle vibration peak in the Y direction and X direction is larger than that in the Z direction. The blasting vibration velocity in all directions continues to decline with the increase of proportional distance as a whole and the decline amplitude of blasting vibration velocity in the X and Y directions is relatively larger than that in the Z direction. When the proportional distance is greater than 50 m/kg$^{1/3}$, and when $\overline{R}$=55.26 m/kg$^{1/3}$, the blasting vibration velocity in the X direction reaches the minimum value of 0.32cm/S, and the blasting vibration velocity in the Y direction reaches the minimum value of 0.51cm/S. Here, although the blasting vibration velocity in the Z direction is 0.16cm/S, its minimum value is 0.08cm/S when $\overline{R}$=19.11 m/kg$^{1/3}$.

When the stratum in the numerical simulation is composed of dolomitic limestone or hard clay, the maximum blasting vibration velocity in the horizontal radial, horizontal tangential and vertical directions of the blasting seismic waves recorded at the monitoring point is obtained, and the variation of the vibration peak with the increase of the proportional distance is obtained, as shown in Figure 5.38 and Figure 5.39.

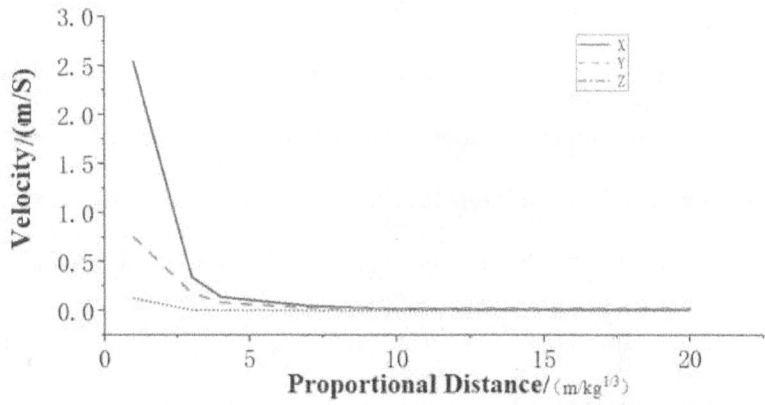

Fig.5.38 Variations of maximum blasting vibration in limestone

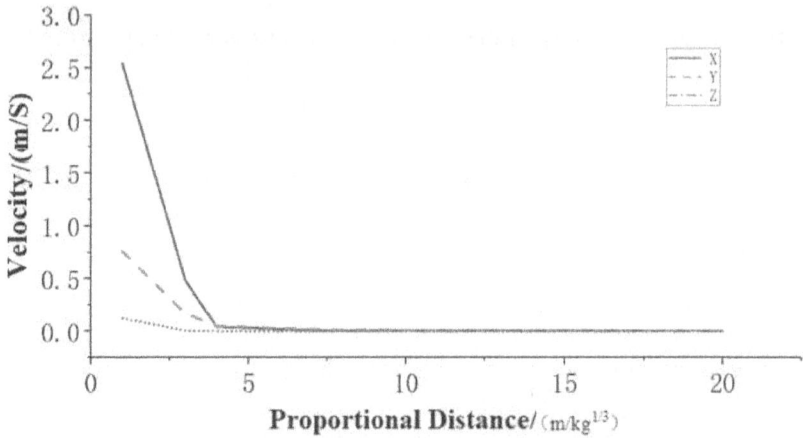

Fig.5.39 Variations of maximum blasting vibration in crushed stone

According to Figure 5.38, in the numerical simulation, when the explosive charge is 25kg and the stratum is composed of weakly weathered dolomitic limestone, the peak vibration values of particles in the X, Y, and Z directions gradually attenuate with the increase of proportional distance. The initial maximum vibration velocity in the X direction is the highest among the three directions, followed by the initial maximum vibration velocity in the Y direction, and the minimum is the initial maximum vibration velocity in the Z direction. The maximum vibration velocity in the X direction attenuates the fastest with the increase of proportional distance, followed by the maximum vibration velocity in the Y direction, and the maximum vibration velocity in the Y direction has the slowest attenuation. After $\overline{R}=10$, all the maximum vibration velocities in the X, Y, and Z directions tend to approach 0.

According to Figure 5.39, in the numerical simulation, when the explosive charge is 25kg and the stratum is composed of gravel soil, the peak vibration values of particles in the X, Y, and Z directions gradually attenuate with the increase of proportional distance. The initial maximum vibration velocity in the X direction is the highest among the three directions, followed by the initial maximum vibration velocity in the Y direction, and the initial maximum vibration velocity in the Z direction is the minimum. The maximum vibration velocity in the X direction attenuates the fastest with the increase of proportional distance, followed by the maximum vibration velocity in the Y direction, and the maximum vibration velocity in the Y direction has the slowest attenuation. However, from the graph and the data in Chapter 4, it can be seen that within the proportional distance range of 3~4 m/kg$^{1/3}$, when the blasting seismic waves just pass through the stratum composed of hard clay, the maximum vibration velocity attenuation velocity in the X direction is further accelerated, resulting in the maximum vibration velocity in the X direction being smaller than that in the Y direction at measurement point #3 at a proportional distance of 4 m/kg$^{1/3}$ when the blasting seismic waves just pass through the stratum composed of hard clay. After the proportional distance is greater than 10 m/kg$^{1/3}$, all the maximum vibration velocities in the X, Y, and Z directions approach 0.

By comparing Figures 5.37, 5.38 and 5.39, it can be seen that both the data calculated using numerical models and the data measured at monitoring points in actual blasting projects show that the maximum vibration velocity in the Z direction is smaller than that in the other two directions, and the maximum vibration velocity attenuation rate in the Z direction is also smaller than that in the other two directions. However, in numerical simulations, it is found that the maximum vibration velocity in the X direction is greater than that in the Y direction. However, in actual engineering measurements, it is found that the maximum vibration velocity in the Y direction is greater than that in the X direction within a proportional distance range of 17~25 m/kg$^{1/3}$. After analysis, it is believed that there are three reasons for this:

(1) The proportion distance of the nearest measuring point in actual engineering is greater than 17 m/kg$^{1/3}$, and the maximum vibration velocity in the X, Y, and Z directions of the near area has not been measured, while the maximum vibration velocity of blasting in the far area is easily affected by many factors.

(2) The detonation method used in the actual engineering and numerical simulation are different. In actual engineering, millisecond delay hole-by-hole detonation is used, while in numerical simulation, centralized detonation is used.

(3) In Figure 5.37, although the maximum vibration velocity in the X direction is greater than that in the Y direction within a proportional distance of 35~50 m/kg$^{1/3}$, the maximum vibration velocity in the X direction is less than that in the Y direction within a proportional distance of 17~25 m/kg$^{1/3}$. In Figure 5.2, the maximum vibration velocity attenuation rate in the X direction is greater than that in the Y direction within a proportional distance of 35~50 m/kg$^{1/3}$. However, in Figure 5.4, when the stratum is composed of hard clay and the proportional distance is 4 m/kg$^{1/3}$ at measuring point #3, the maximum

vibration velocity in the X direction is smaller than that in the Y direction when the blasting seismic waves just pass through the stratum composed of hard clay, which confirms the credibility of the numerical model.

The vibration frequency in the X direction is compared with and verified by engineering examples. MATLAB is used to perform fast Fourier transform on the velocity time-history curves in the X direction of seven measurement points when the stratum is composed of dolomitic limestone and gravel soil. The time-domain signal is converted into a frequency-domain signal to obtain the vibration frequency in the X direction, and Table 5.3 is obtained.

Table 5.3 Data of main blasting vibration frequency in the X direction of numerical simulation

Position		1	2	3	4	5	6	7
Distance (m)		2.92	8.77	11.70	14.62	29.24	43.86	58.48
Proportional dist.		1	3	4	5	10	15	20
limestone	X-Dominant	5.73	14.71	11.21	16.78	4.77	4.06	3.98
crushed stone freq. (Hz)		7.24	15.03	2.15	1.75	2.45	2.90	2.60

According to Table 5.3, the variations of the main frequency of blasting vibration at the seven measuring points with increasing proportional distance when the stratum is composed of weakly weathered dolomitic limestone or gravelly soil, as shown in Figures 5.40 and 5.41.

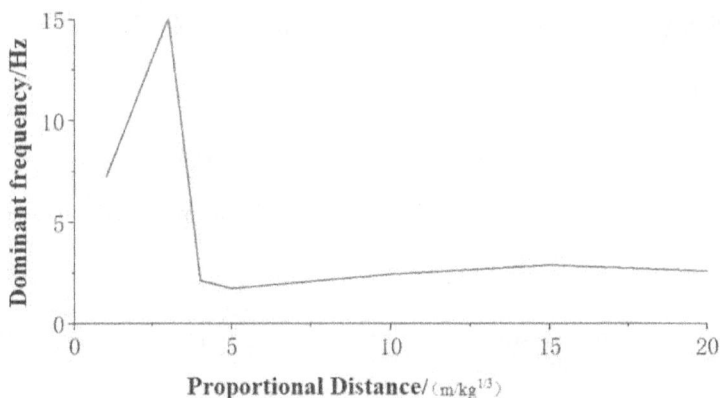

Fig.5.40 Variations of frequency in x direction in dolomitic limestone

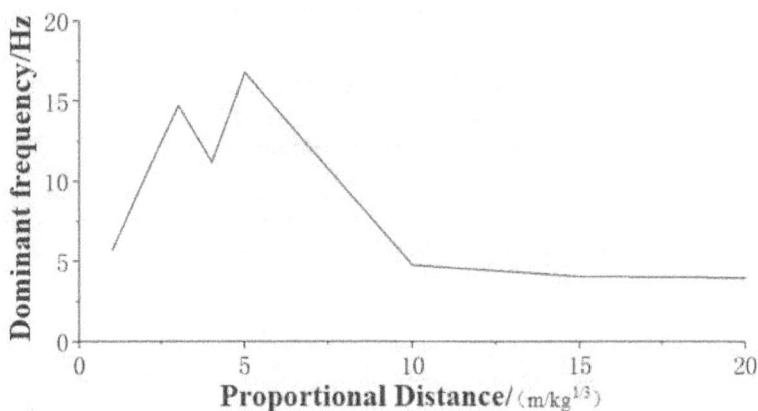

Fig.5.41 Variations of frequency in x direction in crushed stone

From Figures 5.40 and 5.41, it can be seen that overall, the vibration frequency in the X direction increases first and then decreases with the increase of proportional distance. When the stratum is composed of dolomitic limestone, within a proportional distance of 3~4 m/kg$^{1/3}$, that is, when seismic waves cross the stratum, vibration frequency of particles near the stratum in the X direction decreases. At the beginning of crossing the stratum, the vibration frequency in the X direction increases. However, when the proportional distance is within the range of 6~10 m/kg$^{1/3}$, the vibration frequency of particles in the X direction rapidly decreases. When the proportional distance exceeds 10 m/kg$^{1/3}$, the decrease of vibration frequency in the X direction slows down. When the stratum is composed of gravel soil, when the proportional distance is within the range of 3~4 m/kg$^{1/3}$, that is, when the seismic waves cross the stratum, the main frequency of the X direction vibration of particles near the stratum rapidly decreases. When the proportional distance is within the range of 4~5 m/kg$^{1/3}$, that is, when the seismic waves just cross the stratum, the rate of decrease in the X direction vibration main frequency slows down significantly. After the proportional distance exceeds 5 m/kg$^{1/3}$, the main vibration frequency in the X direction has a certain degree of rebound. From the above analysis, it can be concluded that hard strata have a relatively small effect on the main vibration frequency in the X direction. The main vibration frequency in the X direction decreases only when the seismic waves cross the strata. Weak strata have a significant effect on the main vibration frequency in the X direction. When the seismic waves cross the weak strata, the main vibration frequency in the X direction rapidly decreases, and after they cross the weak strata for a certain distance, the main frequency of vibration in the X direction rebounds.

The vibration frequency in the X direction in the engineering example is selected to analyze the blasting vibration frequency through proportional distance to obtain the variations of the vibration frequency with the increase of proportional distance, as shown in Figure 5.7.

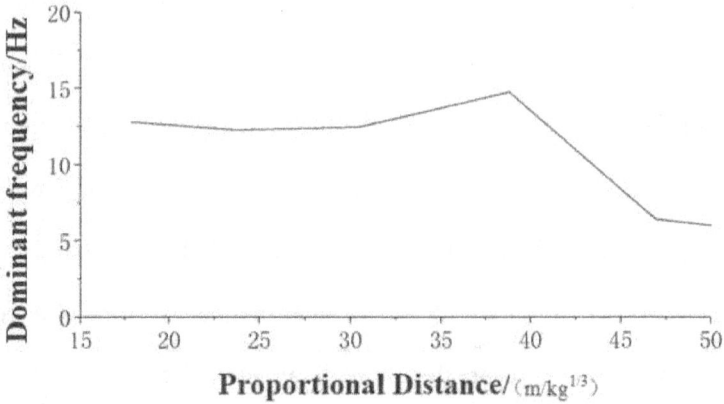

Fig.5.42 Variations of frequency in x direction with proportional

distance

From Figure 5.42, it can be seen that the main frequency of vibration in the X direction first decreases, then increases, and then decreases within a proportional distance range of 17~55 m/kg$^{1/3}$. Based on Figure 5.5, it can be seen that when the stratum is composed of dolomitic limestone, the main frequency of blasting vibration suddenly increases and then rapidly decreases within a proportional distance range of 4~10 m/kg$^{1/3}$, which is consistent with the trend of the main frequency of blasting vibration changing with the increase of proportional distance in the engineering example of Figure 5.7 within a proportional distance range of 35~47 m/kg$^{1/3}$. This proves that when encountering geological inclusions, the main frequency of vibration in the X direction suddenly increases and then rapidly decreases. According to Figure 5.41, the main frequency of blasting vibration in the proportional distance range of 4~15 m/kg$^{1/3}$ first decreases slightly and then slowly increases again with the increase of proportional distance when the stratum is composed of gravel soil. This is consistent with the trend of the main frequency of blasting vibration in the range of 17~38 m/kg$^{1/3}$ with the increase of proportional distance in the engineering example in Figure 5.42. Therefore, it can be proved that when encountering geological inclusions, the main vibration frequency in the X direction first slowly decreases and then slowly increases.

In summary, this numerical simulation can reasonably simulate and explain the propagation process of blasting seismic waves in rock masses with geological inclusions. The numerical simulation is consistent with the actual engineering and has a certain degree of credibility. Based on the blasting parameters used in the mining area and the selection of monitoring points, the final vibration monitoring results have been obtained. The blasting method used in the open-pit mine for engineering observation is millisecond delay hole-by-hole blasting, and the explosives used for blasting are ammonium oil and emulsion explosives. The consumption of explosives used for blasting is $q=0.5~1.1$kg/m^3. Six measuring points with proportional distances of 17.89 m/kg$^{1/3}$, 23.73 m/kg$^{1/3}$, 30.46 m/kg$^{1/3}$, 38.79 m/kg$^{1/3}$, 46.93 m/kg$^{1/3}$ and 55.26 m/kg$^{1/3}$ respectively are selected. Finally, the data obtained from six measurement points are provided separately. The variations of the maximum vibration velocity of blasting seismic waves in the horizontal radial, horizontal tangential, and vertical directions with proportional distances in actual engineering are plotted into maps, which are compared with the results obtained by plotting the maximum vibration velocity of blasting seismic waves in the horizontal radial, horizontal tangential, and vertical directions with proportional distances recorded at monitoring points in numerical simulations when the stratum is composed of dolomitic limestone or gravel soil. This numerical model can be used to simulate the variation law of blasting vibration velocity of blasting seismic waves in the propagation process under different geological materials in actual engineering.

The variations of the main frequency of blasting seismic waves in the X direction with proportional distance in actual engineering are plotted, and fast Fourier transform is performed on the velocity time-history curve of blasting seismic waves in the X direction recorded by monitoring points when the stratum is composed of dolomitic limestone or gravel soil in numerical simulation. The time-domain signal is converted into a frequency-domain signal to obtain the main frequency of blasting in the X direction. The variations of the main frequency of blasting seismic waves in the X direction with proportional distance are plotted. By comparing the two, it is found that the variations of the main frequency of blasting seismic waves in the X direction in the numerical simulation is reflected in practical engineering. Therefore, this numerical model can be used to simulate the variation law of the propagation process of blasting seismic waves under different geological inclusions.

By analyzing and studying the propagation law of blasting seismic waves in rock masses under different strata inclusions under general blasting conditions, displacement time-history curves at measurement points #1~#10 were extracted for granite, dolomitic limestone, gravel soil, and hard clay strata. Within the simulation time range of 0~60MS, the following content and conclusions were drawn:

(1) For measuring points located between the explosion source and the stratum and close to the stratum, if the strength of the rock mass on both sides is greater than or equal to the strength of the stratum, the displacement of the particles at that location will decrease. However, when the strength of the stratum is small, the displacement of the particles at that location will actually increase.

(2) When the blasting seismic waves cross the stratum and the proportional distance is within the range of 3~10m/kg$^{1/3}$, the weak strata can significantly suppress the displacement of particles in the early stage of the explosion, but the growth rate of displacement will be faster after displacement occurs, and the final displacement will increase. The hard strata have a relatively small effect on the final displacement of nearby particles, and also have a relatively small effect on the shape of the displacement time-history curve. Moreover, the maximum displacement is generally smaller than when the stratum is composed of granite.

(3) When the blasting seismic waves cross the stratum and the proportional distance exceeds 15m/kg$^{1/3}$, the weak strata can significantly suppress the displacement of nearby particles in the early stage of the explosion. After the displacement begins to occur, the displacement of the

205

measuring point will increase steadily. The maximum displacement generated is smaller than that when the stratum is composed of granite, and as the proportional distance increases, the maximum displacement generated is smaller compared to that when the stratum is composed of granite. The effect of hard strata on the displacement time-history curve of nearby particles is relatively small.

(4) For measuring points on the same horizontal plane as the explosion source or at the boundary between the right side of the strata and the rock masses above that horizontal plane, when the magnitude of the stratum strength is greater or less than that of the rock masses on both sides, the displacement at that measuring point will increase. When the strength of the stratum is of the same order of magnitude as the rock masses on both sides, it will actually reduce the final displacement at the measuring point. The presence of weak strata will suppress the displacement of the measuring point in the initial stage, and the weaker the strata, the longer the suppression time.

(5) For the measuring point at the boundary between the right side of the strata and the rock masses below the same horizontal plane as the explosion source, regardless of whether the strata are weak or hard, the presence of the formation will increase the final displacement at the measuring point. The presence of weak strata will suppress the displacement of the measuring point in the initial stage, and the weaker the strata, the longer the delay in displacement, and the greater the final displacement at the measuring point.

In addition, regarding the propagation law of blasting seismic waves in rock masses with different strata inclusions under general blasting conditions, the blasting vibration velocities at measuring points #1~#7 are taken when the strata are composed of granite, dolomitic limestone, gravel soil, and hard clay respectively. Through analysis and comparison, the content and conclusions are as follows:

(1) For measuring points #1 and #2 on the left side of the stratum, the vibration velocity curves in the X, Y, and Z directions are generally consistent, and they are basically at the minimum value of the blasting vibration velocity during the entire blasting process in the same time domain. However, for measuring point #2, due to its location between the weak stratum and the blasting source, it is affected by the reflection and transmission of seismic waves due to the existence of the boundary between the rock mass and the stratum. The time-history curve of this measuring point shows an upward trend throughout the entire calculation time. In addition, the reflection and transmission of seismic waves affect the oscillation amplitude of the time-history curve at measurement point #1 after 30MS.

(2) For measurement point #3 on the right side of the stratum, due to its close proximity to the stratum, it is affected by seismic wave reflection and transmission, resulting in a certain upward trend in the time-history curve throughout the entire calculation time. The hard strata have little effect on the vibration waveform of the time-history curve at the measuring point, mainly reflected in the different oscillation amplitudes of the velocity time-history curve after 30MS and the different values of the maximum and minimum

vibration velocities. On the other hand, the weak stratum greatly changes the vibration waveform of the velocity time-history curve at the measuring point. Compared with the vibration waveform of the hard strata, the vibration frequency of seismic waves decreases significantly after passing through the weak strata.

(3) For measuring points #4~# 6 on the right side of the stratum, due to their distance from the stratum, they are minimally affected by seismic wave reflection and transmission. The hard strata have little effect on the vibration waveform of measuring points #4~#6, mainly reflected in the different oscillation amplitudes of the velocity time-history curve after 30MS, as well as the timing of the maximum and minimum vibration velocities. On the other hand, the weak strata greatly change the vibration waveform of the velocity time-history curve of measuring points #4~#7. Compared with the vibration waveform of the hard stratum, the vibration frequency of seismic waves decreases significantly after passing through the weak strata.

(4) For measuring point #7 on the right side of the stratum, due to its distance from the blasting source, both hard and weak strata have little effect on the vibration waveform of this measuring point. This is mainly reflected in the timing of the maximum and minimum values of the blasting vibration velocity throughout the entire blasting process.

(5) Based on the above analysis results, it can be concluded that the geological strata can cause nearby measurement points to experience seismic wave reflection and transmission, resulting in an upward trend in the time-

history curve. The presence of hard strata has little effect on the vibration waveform of the time-history curve of measuring points within a proportional distance range of $7\sim15$m/kg$^{1/3}$. This is mainly reflected in the different oscillation amplitudes of the velocity time-history curve after 30MS as well as the different timing of the maximum and minimum values of vibration velocities. However, weak strata greatly change the vibration waveform of the velocity time-history curve of measuring points within this proportional distance range, and the vibration frequency of seismic waves decreases significantly after passing through weak strata. After the proportional distance exceeds 20m/kg$^{1/3}$, both hard and weak strata have little effect on the vibration waveform of the velocity time-history curve.

(6) In the case of using 25kg of explosives, the maximum blasting vibration velocity curves in the X, Y, and Z directions at each measuring point under different geological inclusions are basically the same. Only within a proportional distance of $3\sim8$m/kg$^{1/3}$ will there be an effect, which means that only near the strata will there be a considerable effect on the maximum blasting vibration velocity in the X direction. Due to the large quantities of explosives used, the blasting vibrations in the X, Y, and Z directions in the vicinity are mainly directed away from the blasting source, resulting in a smaller minimum vibration velocity in the blasting vicinity. As a result, the curve of the minimum blasting velocity in the Y direction is not a decreasing curve. The X direction mainly refers to the minimum blasting velocity after 3m/kg$^{1/3}$. However, the curve of the minimum blasting velocity in the Z

direction is roughly the same. The four curves of the minimum blasting velocity in the Y direction all have an upward first and then downward process. When the stratum is composed of gravel soil or hard clay, this trend occurs before the seismic waves cross the stratum, and when the stratum is composed of gravel soil or hard clay, this trend occurs after the seismic waves cross the stratum. In summary, the presence of strata has a relatively small effect on the maximum value of the blasting vibration velocity in the three directions. In the case of weak strata, the attenuation amplitude of the minimum blasting vibration velocity is faster than in the case of hard strata.

6 Vibration of Pre-splitting Blasting under Joint Conditions

6.1 Rock and Joint Material Model

Due to the long-term weathering and erosion, the edge rock mass also contains a large number of joint fissures and fractures inside. As a result, the rock material is usually characterized by discontinuity, heterogeneity, and anisotropy. At present, it is still difficult for us to fully simulate the various structural planes such as fissures and fractures in the rock mass. For the artificial control over the number, distribution, and properties of various structural planes exerted by researchers in the process, the numerical simulation cannot fully reflect the structural characteristics of the rock mass. To make the simulation analysis of the blasting numerical model of joint rock mass more convenient and operable, researchers often adopt simplified rock material models while ensuring that the numerical simulation is as consistent as possible with the actual engineering.

LS-DYNA is adopted to simulate the pre-splitting and forming of slope rock mass under different joint occurrence conditions at different angles. To ensure the reliability of the simulation process and calculation results of the pre-splitting and forming, the rock material in the numerical simulation model is simulated by utilizing the concrete material model *MAT_JOHNSON_HOLMQUIE_CRITE (H-J-C model), whose main parameters are shown in Table 6.1.

Tab. 6.1 Parameters of rock material model

density (g/cm³)	Elastic modulus (GPa)	Poisson's ratio	shear modulus (GPa)	Compressive strength (MPa)	Tensile strength (MPa)
2.1	0.1486	0.1	9.00	0.00048	0.0004

Normalized tensile strength A	Normalized pressure hardening coefficient B	Strain rate coefficient C	Damage constant D1	Damage constant D2	Plastic strain
0.79	1.60	0.007	0.04	1.0	0.01

The joint material mainly adopts the plastic dynamic hardening material model MAT-PLASTIC_KINEMATIC commonly used in LS-DYNA software, and the main parameters are shown in Table 6.2.

Tab. 6.2 Parameters of the joint material model

Density (g/cm³)	Elastic modulus (GPa)	Poisson's ratio	yield stress (GPa)	shear modulus (GPa)	hardening parameter
1.86	10	0.3	0.40	9.00	1.00

At present, there are no perfect models for domestic and foreign researchers to describe the complex constitutive relationships of rock media accurately. After repeated verification experiments, some models have been gradually recognized and widely applied in engineering because they have some important reference value in theory and practice and conform to some characteristics of relevant engineering materials.

In 1864, Tresca proposed a hypothesis about yield conditions, stating that the reason why materials yield is that the maximum shear stress reaches a corresponding limit value, when the Tresca yield conditions can be expressed by a simple formula:

$$\tau_{max} = k \tag{6.1}$$

If $\sigma_1 > \sigma_2 > \sigma_3$, the conditions for the establishment of (6.1) can be expressed as:

$$\begin{cases} \sigma_1 - \sigma_2 = \pm 2k \\ \sigma_2 - \sigma_3 = \pm 2k \\ \sigma_1 - \sigma_3 = \pm 2k \end{cases} \tag{6.2}$$

Equation (6.2) can be transformed into an expression for a general stress function:

$$[(\sigma_1 - \sigma_2)^2 - 4k^2][(\sigma_2 - \sigma_3)^2 - 4k^2][(\sigma_1 - \sigma_3)^2 - 4k^2] = 0 \tag{6.3}$$

Tresca yield conditions are widely accepted because the yield conditions, when the principal stress is known, are the linear function of the principal stress, which facilitates solution. However, the shortcomings, such as the presence of corner points on the yield condition curve and the neglect of the influence of intermediate principal stresses, pose difficulties in mathematical calculations.

Recognizing the inadequacy of Tresca yield conditions, Misses proposed a hypothesis about the distortion energy condition as the yield condition in 1993. Assuming that when a point in an object yields, the distortion energy corresponding to its stress state reaches a corresponding limit value k, the distortion energy condition of that point can be expressed by the following formula:

$$J_2 = GS_{ij}e_{ij} = k^2 \tag{6.4}$$

Where
$$S_{ij} = \sigma_{ij} - \sigma_m \delta_{ij};$$
$$e_{ij} = \frac{S_{ij}}{2G};$$

k is parameters representing the yield characteristics of materials. For the convenience of using it in plastic mechanics, formula (6.4) can be expressed in the form of equivalent stress:

$$\sigma_e = \frac{1}{\sqrt{2}}\sqrt{(\sigma_1 - \sigma_2)^2 + (\sigma_2 - \sigma_3)^2 + (\sigma_1 - \sigma_3)^2} = k^2 \tag{6.5}$$

and

$$\sigma_e = \sqrt{3J_2} = \sqrt{3}k \tag{6.6}$$

When determined by simple tensile testing $k = \frac{1}{\sqrt{3}}\sigma_e$,

$$\sigma_e = \sigma_s \tag{6.6}$$

Therefore, the Misses yield condition can be expressed as: when the equivalent stress reaches the corresponding yield limit of simple tension, the material gradually enters the plastic state. By using Misses equivalent stress, this study simulates and analyzes the fragmentation characteristics as well as the pre-splitting and forming effect of rock mass produced by pre-splitting blasting under joint occurrence conditions.

In the process of simulating slope pre-splitting blasting by using ANSYS/LS-DYNA, to better simulate the entire explosion process of explosives, the stable detonation wave front after the detonation is stabilized should meet the following conditions:

$$\rho_1 = \frac{k+1}{k}\rho_2 \tag{6.7}$$

$$u = \frac{k+1}{k}D \tag{6.8}$$

$$C = \frac{k}{k+1}D \tag{6.9}$$

$$P = \frac{1}{k+1}\rho D^2 \tag{6.10}$$

In the formula, ρ_1 is the pressure of the detonation products on the detonation wave front; ρ_2 is the density of explosives; μ is the velocity of particles; C is the speed of sound; k is the multi-party index; D is the detonation velocity of explosives.

The mass equation of the motion equation of detonation products can be expressed as:

$$\frac{\partial \rho}{\partial t} + \nabla \cdot (\rho \mu) = 0 \tag{6.11}$$

The momentum equation of the motion equation of detonation products can be expressed as:

215

$$\frac{\partial \mu}{\partial t} + (\mu \cdot \nabla) = -\frac{\nabla_p}{\rho} \tag{6.12}$$

The energy equation of the motion equation of detonation products can be expressed as:

$$\frac{\partial}{\partial t}\left[\rho\left(e + \frac{\mu^2}{2}\right)\right] = -\nabla\left[\rho\mu\left(\rho\mu + \frac{P}{\rho} + \frac{\mu^2}{2}\right)\right] \tag{6.13}$$

In blasting simulation, due to the large pressure fluctuation range of the so-called detonation products generated by the explosion of high-energy explosives, it is difficult to find the state equations that can fully and accurately reflect all pressure fluctuation ranges in the simulation in LS-DYNA. The most commonly used and effective one is the JWL (Jones-Wilkens-Lee) state equation, which can accurately describe the expansion process of the explosive gas and the changes of chemical energy during the explosion of explosives. Generally, the JWL state equation can be expressed in the form of

$$P = A\left(1 - \frac{\omega}{R_1 V}\right)e^{-R_1 V} + B\left(1 - \frac{\omega}{R_2 V}\right)e^{-R_2 V} + \frac{\omega E_0}{V} \tag{6.14}$$

where V is the relative volume; E_0 is the initial internal energy density; A, B, R_1, R_2, and ω are independent physical constants that describe the JWL equation.

To better simulate the effect of joint occurrence conditions at different angles on the formation effect of pre-splitting blasting, the explosive model used in the blasting model constructed in this study is the high-performance explosive material model *MAT_HIGH_EXPLOSIVE-BURE and the corresponding state equation JML in the LS-DYNA software. The main parameters are shown in Table 6.3.

Tab. 6.3 Parameters of explosives and state equation

Density (g/cm³)	Detonation velocity (m/S)	Explosive pressure (GPa)	E (MJ•m⁻³)	A (GPa)	B (GPa)	R₁	R₂	ω
1.30	3200	316	4192	214.4	18.2	4.2	1.0	0.15

In LS-DYNA software, there is a specifically designed for air simulation, together with an ideal gas state equation *EOD_LINEAR_POLYNOMIAL with corresponding linearly-distributed internal energy. The *NULL material model can simulate air material, whose pressure can be represented by the following equation:

$$\begin{cases} p = C_0 + C_1\mu + C_2\mu^2 + C_3\mu^3 + (C_4 + C_5\mu + C_6\mu^2)E \\ \mu = \dfrac{\rho}{\rho_0} - 1 \end{cases} \tag{6.15}$$

In the formula, P is the air pressure; μ is the specific volume; ρ is the air quality density; ρ_0 is the reference mass density, $C_0 - C_6$ is the coefficient of the seven air state equations; E is the initial energy density.

The air model parameters and state equation parameters in the blasting model constructed in this study are shown in Table 6.4.

Tab. 6.4 Parameters of air and state equations

ρ(g/cm³)	C₀	C₁	C₂	C₃	C₄	C₅	C₆	E (MPa)	V₀
1.29E-5	0	0	0	0	0.4	0.4	0	2.5E-4	1

In the ANSYS/LS-DYNA programs, Lagrange algorithm, Euler

algorithm, and ALE algorithm are mainly used to handle explosion numerical simulation. For the physical process in the same explosion numerical simulation, the algorithm methods of A, B, and C can be used to describe the deformation of the object and analyze their differences, as shown in Figure 6.3.

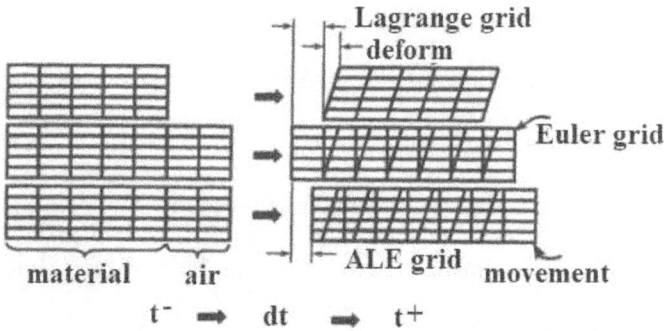

Fig. 6.3 Explosion numerical simulation process

Based on the coordinates of matter, the Lagrange algorithm mainly focuses on the stress-strain analysis of solid material structures and can accurately describe the relevant movements of the material structure boundaries. As shown in Figure 6.3, the grid elements are attached to the material, but they allow the contact, separation, sliding and rubbing of the active and passive surfaces. However, this algorithm is not perfect enough and has the disadvantage of not being able to analyze large deformation problems. When dealing with blasting problems, the grid is prone to large deformation and serious distortion, which may easily lead to calculation termination, producing adverse effects on the calculation results.

Different from Lagrange algorithm, the Euler algorithm mainly solves problems related to extreme deformation based on spatial coordinates. Since the material structure and the divided grid are independent of each other, the calculation of the Euler algorithm is equivalent to overlapping calculations of two layers of grids. The grid's shape, size, and spatial position remain unchanged throughout the analysis process, and it can be considered that the material flows in the grid. The fixed accuracy of numerical calculation in iterations results in the flowing phenomenon of materials between grids, making the capture of material boundaries difficult and the calculations time-consuming.

Unlike Lagrange algorithm and Euler algorithm, which have their own advantages and disadvantages in blasting simulation, the ALE algorithm can be said to have all the advantages of the above two algorithms and can address all their shortcomings in simulation. The ALE algorithm not only absorbs the obvious advantages of Lagrange algorithm in dealing with structural boundary motion and other related problems, being able to track the motion of material structure boundaries effectively in simulations, but it also incorporates the advantages of Euler algorithm in dealing with internal mesh division and other related problems, allowing internal mesh elements to exist independently of material entities. Its accuracy in dealing with large deformation problems is also higher than that of Euler algorithm. In blasting simulations, explosives and air are usually defined as ALE fluid elements, while the structure of the blasted material is defined as Lagrange elements, which define explosives and structural elements through fluid structure coupling. However, this method requires high parameter accuracy, otherwise abnormal phenomena such as negative volume and infinite node velocity may occur, leading to simulation distortion and prevents the calculation from proceeding normally.

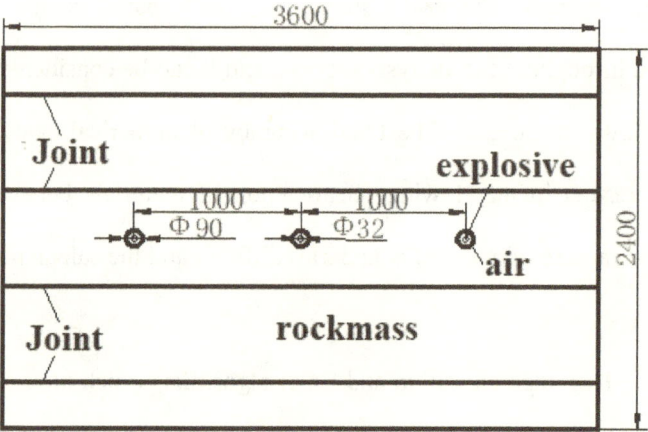

(a) When α is 0°

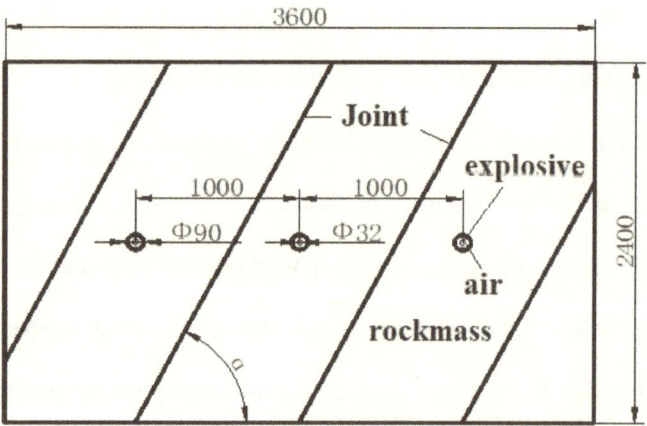

(b) When α is 0 °, 30 °, 45 °, 60 °, 75 °, and 90 ° respectively

Fig. 6.4 Top views of computing model (unit: mm)

Here, LS-DYNA software was used to simulate the effect of occurrence conditions at different angles on the formation of pre-splitting blasting of rock mass. Six sets of related models are established to study the formation effect of pre-splitting blasting in joint rock mass. In each set of models, there are three blast holes with a diameter of 90mm and a spacing of 100cm. A strip-shaped 2# rock emulsion explosive model roll with a diameter of 32mm is placed in the middle of the blast holes, while an air model is placed between the roll and the blast hole wall. The joint rock mass model is 360cm long and 240cm wide. For the convenience of study, it is assumed that there is one set of parallel joints inside the rock layer with a joint structure plane thickness of 1cm that run through the entire model. The six sets of joint rock mass models with occurrence conditions at different angles are constructed, and the angles α between the joints and the boreholes are 0°, 30°, 45°, 60°, 75°, and 90° respectively. The specific calculation models are shown in Figure 6.4.

All the four sides of the six sets of pre-splitting blasting models of joint rock mass with different angles as non-reflective boundary conditions. The unit grids of the joint rock material are divided by using the Lagrange grid elements division method, while the unit grids of explosive materials and air materials are divided by adopting the multi-material ALE elements. Therefore, the fluid-solid coupling algorithm is used in this study to simulate and calculate the rock mass, joints, air, and explosives in different angle joint models.

In addition, the detonation of explosives is simulated by using the method of central-point initiation, and the termination time of the blasting simulation calculation in this model is set to 1000 µS.

6.2 Pre-splitting Simulation of Joints at Different Angles

For rock masses with joint occurrence conditions, the explosive stress waves generated by explosive explosions during pre-splitting blasting are prone to transmission and reflection at weak structural surfaces like joints. After the explosive stress waves pass through the joint structural surfaces, the explosive energy they carry will decrease to some extent, but it will continue to do work on the joint rock mass and further promote the penetration and formation of cracks. However, under normal circumstances, the propagation direction of cracks generated by pre-splitting blasting is related to the occurrence angle of joints. Therefore, the transmission phenomenon of stress waves in pre-splitting blasting simulation is also an important focus of this study.

Due to the fact that the tensile strength, cohesion, and other related parameters of the joint material selected in the simulation will be much smaller than the rock compressive strength and other related parameters selected in the simulation, the explosive stress waves generated by the explosion will first compress the borehole wall of the rock mass at the beginning of the blasting stage. When the ensile stress generated in the joint model material under the explosion effect of the explosive model material exceeds the tensile strength of the joint rock material itself, it will inevitably lead to the trend of explosive cracks in the joint rock mass initiate along the joint direction and further develop towards this trend.

To further investigate the blocking effect of joints on the propagation of stress waves in rock masses with joint occurrence conditions at different angles, a series of nodes can be set up along the direction of the borehole connection as monitoring points. Since the inclined joints in the model pass through the midpoint of the line connecting two boreholes, joint survey lines can be arranged along the lines connecting the two boreholes in the model, as shown in Figure 6.5.

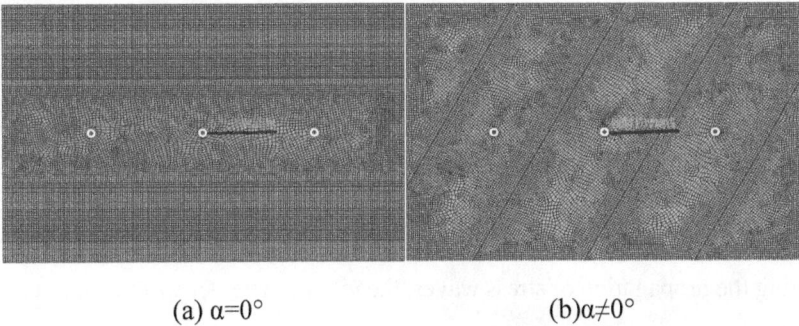

(a) α=0°   (b)α≠0°

Fig.6.5 Monitoring points for structures with different angles

To obtain the vibration velocity of each monitoring point, this study arranges approximately 15 nodes equidistant from the joint measurement line positions marked in the model as vibration monitoring points, and uses LS-DYNA for pre-splitting blasting simulation calculation. Based on the simulation calculation results, the vibration velocity of each monitoring node in the model can be obtained through post-processing, as shown in Figure 6.6.

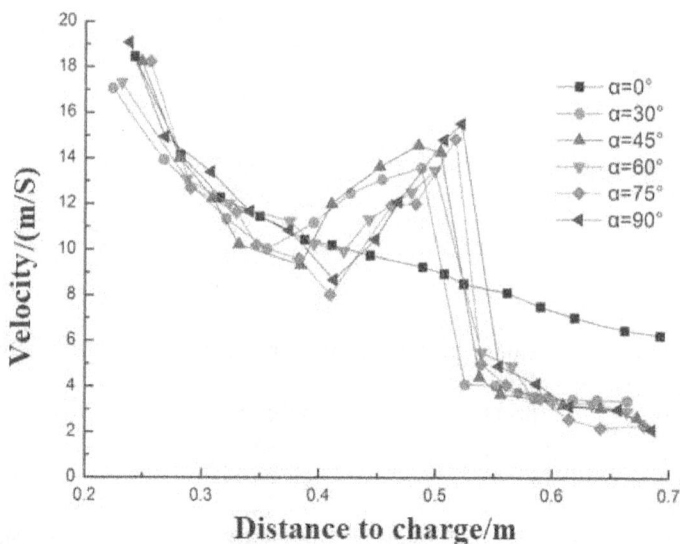

Fig. 6.6 Vibration velocity of different angle structural surface model

From Figure 6.6, the vibration velocity trend of the measuring points in the joint model at different angles can be clearly observed. Due to the existence of joints, the explosion stress waves will produce obvious reflection when propagating to joint structure surfaces in the rock mass. Generally, during the propagation of stress waves, the vibration velocity of the measuring points near the joints will increase significantly, and then decrease significantly in an instant. Because the joint absorbs most of the energy during the propagation of stress waves and significant reflection occurs here, the energy decreases significantly after passing through the joint, leading to a significant decrease in the vibration velocity of the measuring point behind the joint.

By comparing the vibration velocity of measuring points on the joint model measurement line with joint occurrence conditions at different angles, it can be found that the vibration velocity of measuring points behind the joint increases with the increase of joint angles. This indicates that as the angle α between the joint and the borehole increases, the energy that can pass through the joint will also increase. Therefore, it can also be considered that the transmissivity of the joint will become stronger as the value of angle α between the joint and the borehole increases.

Due to the complex form of the transmission-reflection relationship of stress waves passing through joints and the difficulty in obtaining parameters, a common practice is to use the ratio of the amplitudes of nodes at the same position with and without joints as the transmittance:

$$T = \frac{A_{joint}}{A_{intact}} \tag{6.16}$$

In the formula, T is the transmittance; A_{joint} is the amplitude of a node with the presence of joints; A_{intact} is the amplitude of the node without joints.

During the calculation process, the joint rock mass model with joint occurrence conditions at different angles is used to set several nodes as monitoring points at the same position on the line connecting boreholes and ensure that the positions of the measuring points are basically the same. Then, the amplitude data of a certain node with and without joints are collected separately. To avoid stress superposition in simulation calculations, a strategy of detonating only the middle borehole is adopted, which can obtain the transmittance of different angles, as shown in Figure 6.7.

From Figure 6.7, it can be clearly observed that the transmittance of the joint will increase with the increase of the value of angle α between the joint and the borehole. In the constructed joint rock mass model, the joint with a connection angle of $\alpha=90°$ between the joint and the borehole has the highest transmittance, and the joint with a connection angle of $\alpha=30°$ between the joint and the borehole has the lowest transmittance. This situation indicates that as the angle α between the joint and the borehole increases, the energy of the stress wave passing through the joint will also increase. The tensile stress experienced by the joint in the direction of the borehole connection will also increase at the same time, indicating that the probability of rock cracking along the borehole connection will increase as the angle α between the joint and the borehole connection direction increases.

Using ANSYS/LS-DYNA program to simulate the pre-splitting blasting and forming effect of rock mass with joint occurrence conditions at different angles, this study establishes six joint rock mass models with different angles, where the angle α between the joints and the line connecting the blastholes is of 0°, 30°, 45°, 60°, 75°, and 90° for pre-splitting blasting and forming numerical simulation. During the numerical simulation of the pre-splitting blasting and forming process, screenshots are taken every 200 μ s, as shown in Figures 6.8 to 6.13.

As shown in Figures 6.8 to 6.9, we can clearly see the entire simulation process of the complete evolution and formation of cracks around the borehole

226

under the combined action of explosive stress and explosive gas pressure in the joint rock mass model under joint occurrence conditions at different angles during pre-splitting blasting.

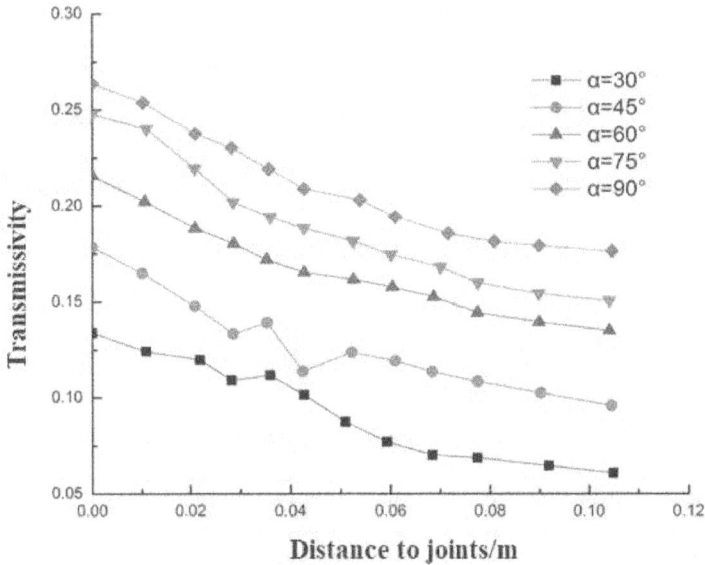

Fig. 6.7 Transmissivity in structural surface models with different angles

When t=0 µS, the explosives has not started to explode. There are four parallel joint lines in the joint model. In the model with the angle α=0° between the joint and line connecting the boreholes, none of the joint lines is set at the position of the line connecting boreholes; instead, they are set on both sides of the line. The joint lines in the other five sets of joint models all pass through the midpoint between the boreholes.

When t=200 µS, a small crushing zone is formed around the borehole wall under the action of explosive stress waves, and there are some randomly distributed radial microcracks outside the crushing zone. Subsequently, as the stress wave continues to propagate, the radial microcrack zone gradually expands deep into the surrounding rock of the borehole wall, forming several less obvious radial main cracks.

When t=600 μS, the blasting model of joint rock mass is basically in the stage of explosive gas pressure. At this time, the gas generated by the explosion will wedge into the previously formed crushing zone and radial crack zone, and at the same time, it may cause relatively large tensile stress at the crack tip, which will further promote the expansion of the radial main crack.

When t=1000 μS, the explosive cracks generated between the blast holes under the action of blasting gas can basically achieve through formation. By observing the forming effect of joint rock mass models under joint occurrence conditions at different angles, and by analyzing and comparing the characteristics of pre-splitting blasting and forming of joint rock mass models under joint occurrence conditions at different angles, we can draw corresponding conclusions.

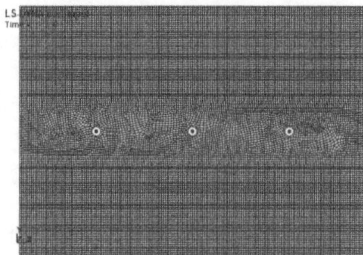 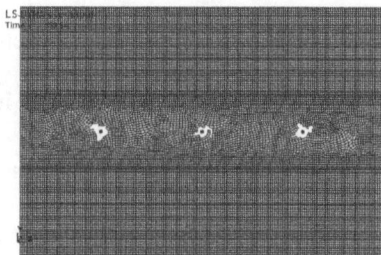

t=0μS   t=200μS

t=400μS   t=600μS

t=800μS   t=1000μS

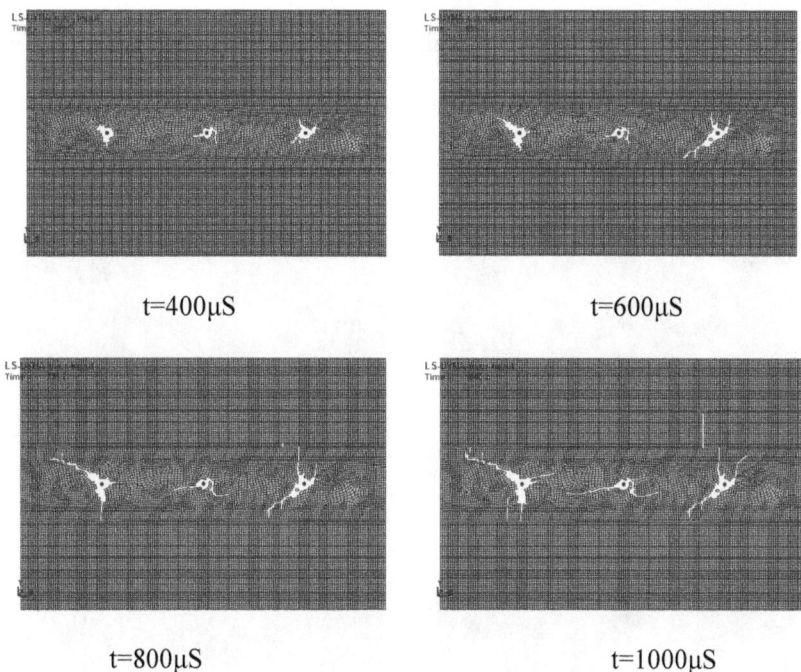

Fig. 6.8 Pre-splitting forming when α=0°

From Figure 6.8, it can be clearly observed that when the angle α=0° between the joint and the borehole, a crushing zone and a radial microcrack zone are formed in a small area around the borehole under the action of explosive stress waves. Then, under the action of explosive gas pressure, further expansion occurs along the radial microcrack direction. The microcracks on both sides of the borehole gradually expand towards the joint, and the horizontal microcracks rapidly expand along the direction of the parallel joint and the joint, ultimately achieving the penetration of pre-splitting cracks and the effect of pre-splitting formation. The joints in the model will experience the failure of joint units in the numerical simulation of rock pre-splitting blasting, and have a certain guiding effect on the generation of blasting cracks, which can fully reproduce the process of pre-splitting blasting and formation of joint rock mass under blasting load.

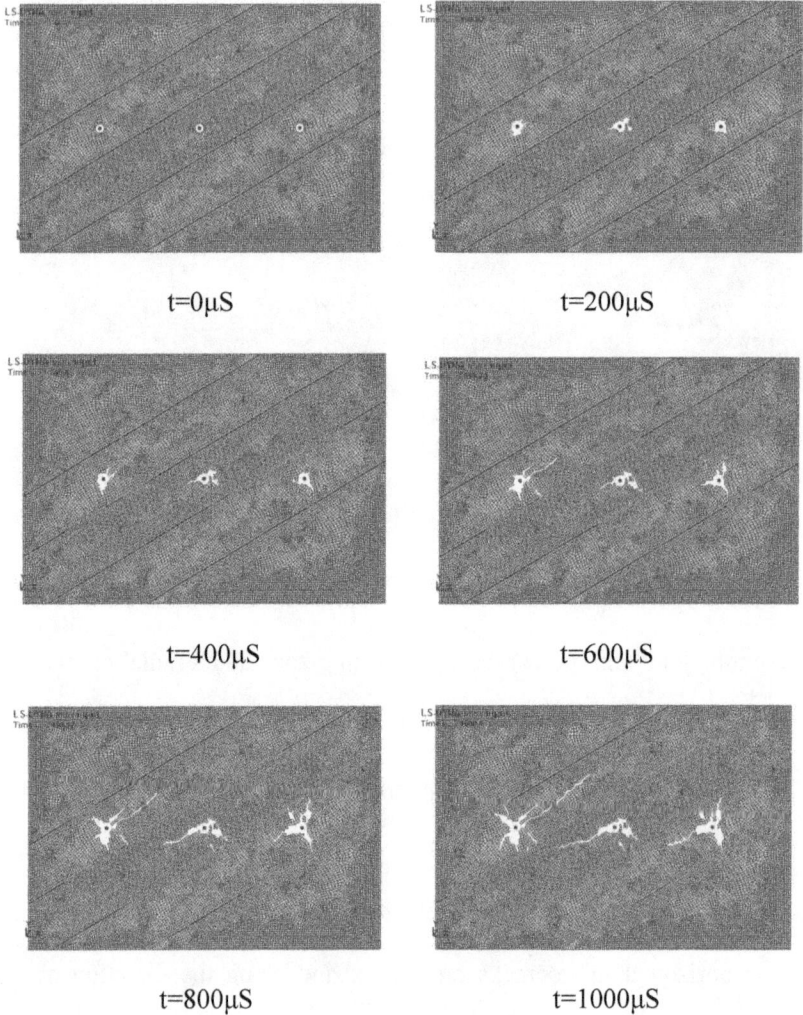

t=0μS

t=200μS

t=400μS

t=600μS

t=800μS

t=1000μS

Fig. 6.9 Pre-splitting molding when α=30°

As clearly observed in Figure 6.9, when a joint exists between two blasting holes with an angle α=30° relative to the line connecting the holes at an angle of, the failure process under the action of explosive stress waves is similar to that in Figure 6.8. A small crushing zone and radial microcracks are formed in the initial stage.

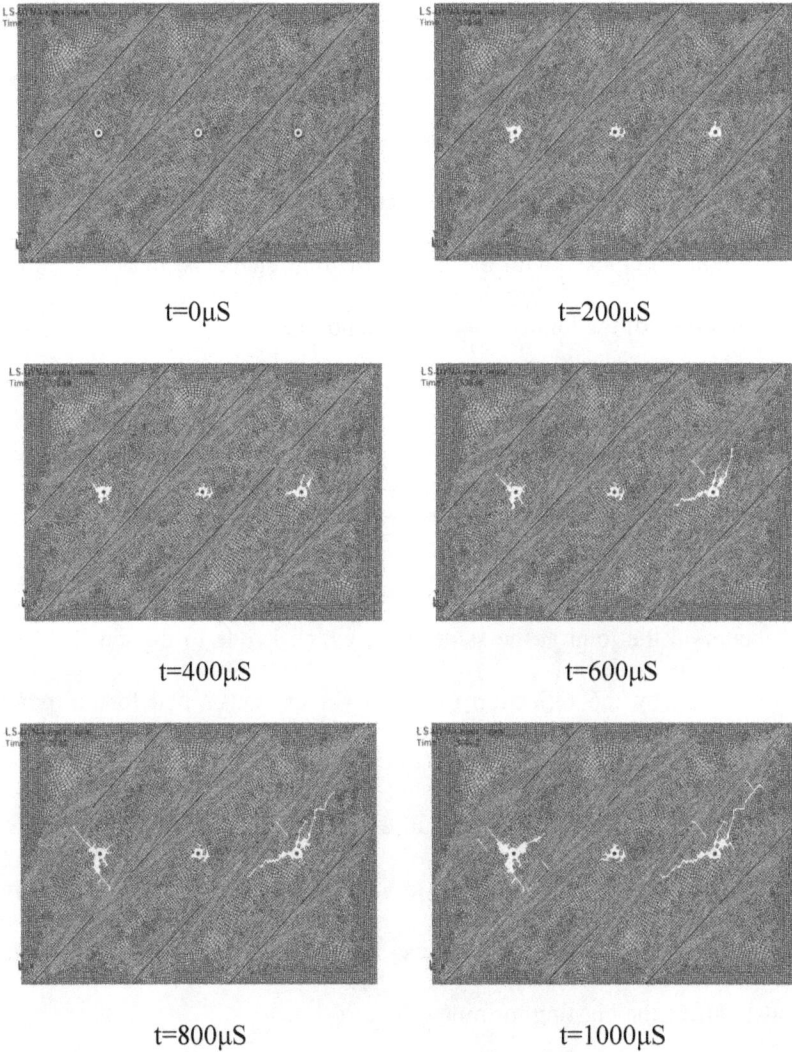

t=0μS   t=200μS

t=400μS   t=600μS

t=800μS   t=1000μS

Fig.6.10 Presplitting forming when α=45°

Subsequently, under the pressure of the explosive gas, further propagation occurs along the radial direction of microcracks, with horizontal microcracks gradually expanding towards inclined joints. These horizontal microcracks then rapidly propagate along the direction parallel to the joint and eventually reach the joint itself, achieving the effect of pre-cracking penetration and pre-splitting formation. As for the main cracks in other directions, there is no significant expansion. It can also be seen in Figure 6.9 that the joint in the model plays a noticeable guiding role in the propagation of blast-induced cracks. With the failure of joint elements, the pre-splitting formation effect of the jointed rock mass is not ideal.

As can be seen in Figure 6.10, when there a joint exists between two boreholes at an angle of $\alpha=45°$ relative to the line connecting the holes, the failure process under explosive stress waves is similar to that in Figure 6.8. Several radial main cracks form around the borehole wall, but local damage also occurs on the joint at the same time, which is due to the tensile stress failure caused by the reflection of stress waves transmitted to this point. Subsequently, under the pressure of the blast-induced gas, the stress concentration effect at the joint end caused the blast-induced cracks to deflect and propagate towards the joint end to varying degrees, ultimately resulting in the penetration failure of the cracks between the two boreholes, which seriously affect the blasting forming effect.

As can be seen in Figure 6.11, when joint exists between two boreholes at an angle $\alpha=60°$ relative to the line connecting the holes, the forming effect is similar to that in Figure 6.10.

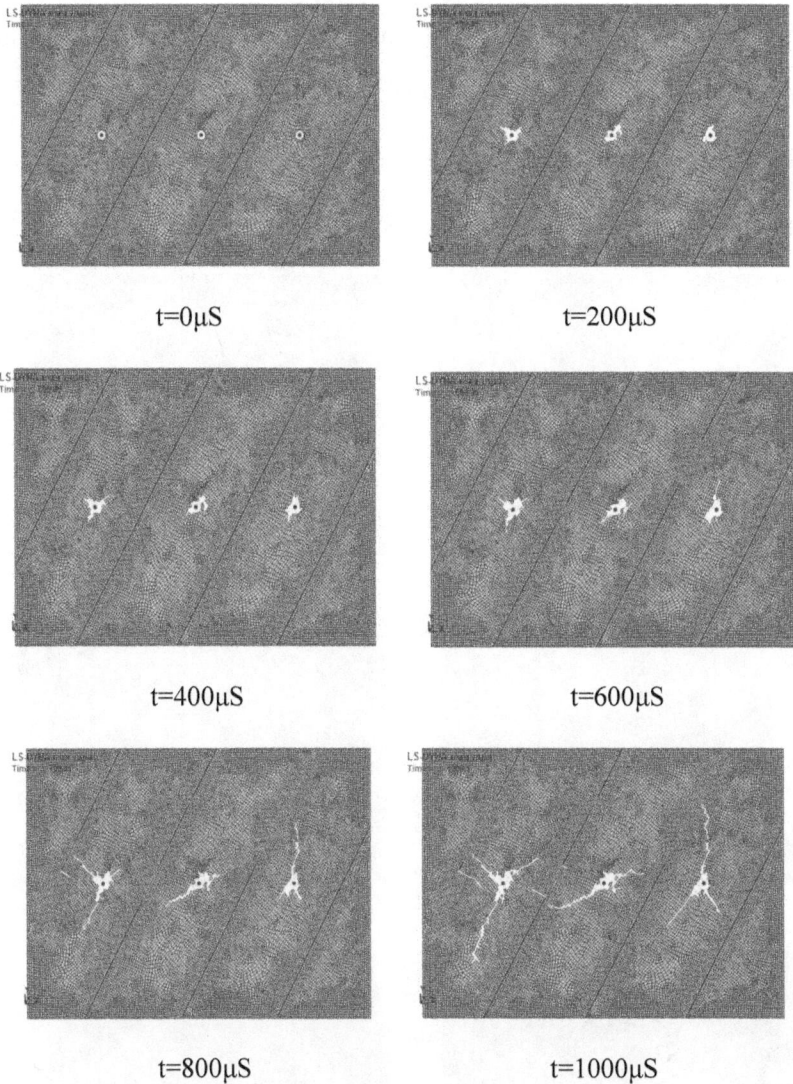

t=0μS

t=200μS

t=400μS

t=600μS

t=800μS

t=1000μS

Fig. 6.11 Pre-splitting forming when α=60°

Cracks closer to the joint surface clearly exhibits a tendency to deflect and propagate towards the end of the joint. Due to the blocking effect of the joint surface, the original main cracks fail to further propagate in their initial direction and multiple "secondary branches" of blast-induced main crack

233

appear, resulting in severe damage to the local rock material. Evidently, the joint still have a significant promoting effect on the propagation of blast-induced cracks in jointed rock masses, but the pre-splitting blasting formation effect of jointed rock mass remains compromised.

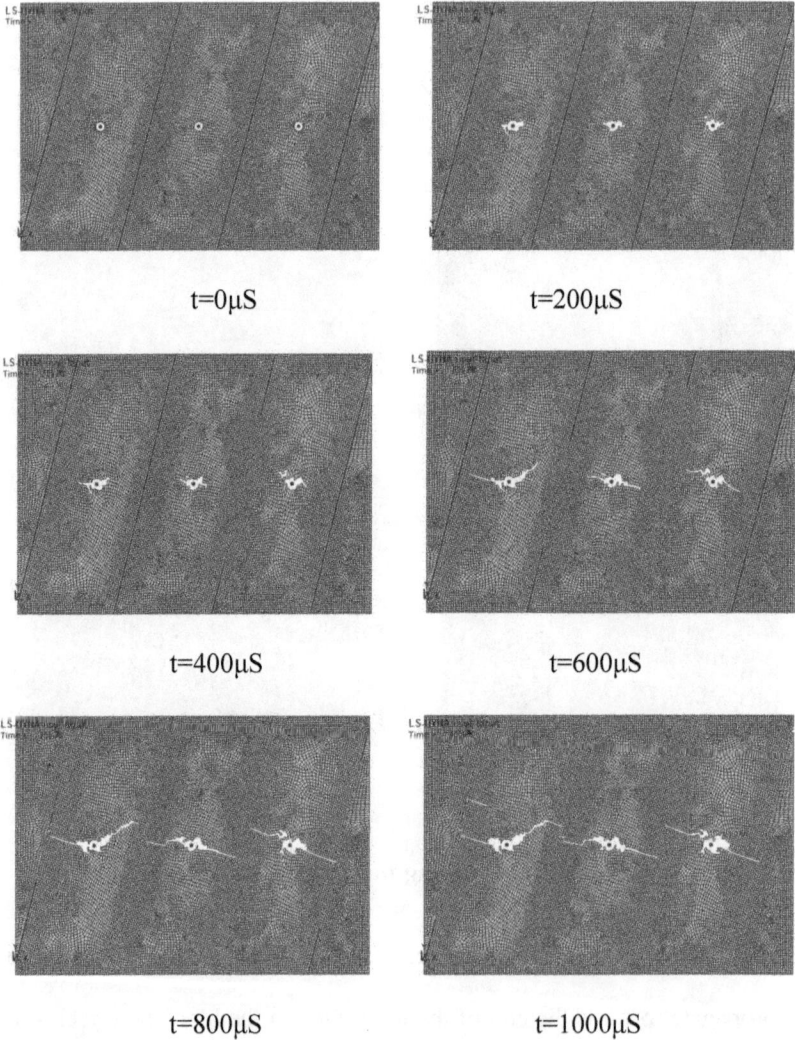

t=0μS

t=200μS

t=400μS

t=600μS

t=800μS

t=1000μS

Fig. 6.12 Pre-splitting forming when α=75°

As can be seen in Figure 6.12, when a joint exists between two boreholes at an angle of α=75° relative to the line connecting the holes, the initial response under explosive stress waves involves the formation of several radial main cracks around the borehole wall, accompanied by localized damage along the joint. Subsequently, under the pressure of the blast-induced gas, the stress concentration at the joint end causes the blast-induced cracks to deflect slightly and propagate towards the joint end to varying degrees. In this scenario, the blocking effect of the joint surface on crack propagation is no longer significant, ultimately allowing horizontal blast-induced cracks to penetrate smoothly. The joint plays a certain guiding role on crack propagation, and the blast-induced cracks exhibit a tendency to extend or deflect towards joints in the process of propagating along the original direction, but penetration is ultimately achieved. The formed surface has obvious serrated characteristics.

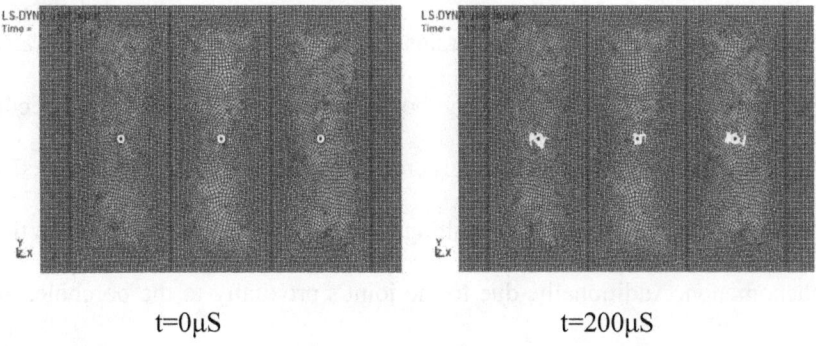

t=0μS   t=200μS

t=400μS　　　　　　　　　　　t=600μS

t=800μS　　　　　　　　　　　t=1000μS

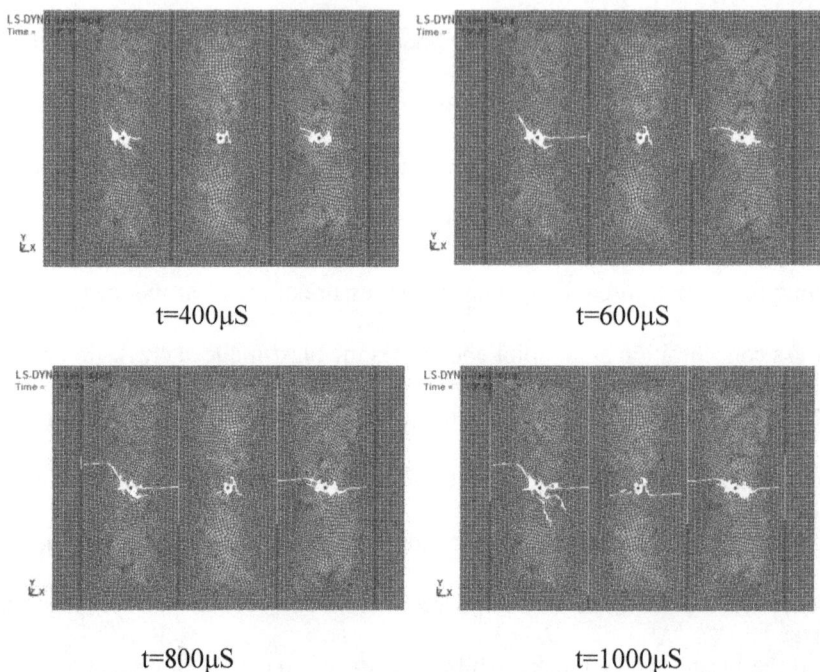

Fig. 6.13 Pre-splitting forming when α=90°

Compared with the pre-splitting blasting effect for a jointed rock mass
with an angle α=75° between the joint and the line connecting boreholes in
Figure 6.12, Figure 6.13 demonstrates that when the angle α=90°, the cracks
extend vertically outward from the boreholes towards the joint. This occurs
because the joint most effectively reflects stress waves in the direction
perpendicular to its plane, resulting in the most pronounced reflection
phenomenon. Additionally, due to the joint's proximity to the borehole, the
intensity of stress waves acting on it is maximized. Consequently, the cracks
exhibit a predominant tendency to develop in the direction perpendicular to
the joint, ultimately achieving the formation of the fracture zone of the

damaged joint.

In summary, by comparing the pre-splitting processes and outcomes of the jointed rock mass models under joint occurrence conditions at six different angles, it can be clearly seen that the cracks generated by pre-splitting blasting tend to be more flat and easier to penetrate as the angle α between the joint and the line connecting the boreholes increases. As for the pre-splitting forming effect of the jointed rock mass models with joint occurrence conditions at six different angles, it is evident that the pre-splitting forming effect is the worst when the angle α is 45° whereas the pre-splitting forming effect is the best when the angle is 90°, with no serrated phenomena and relatively flat penetrating cracks.

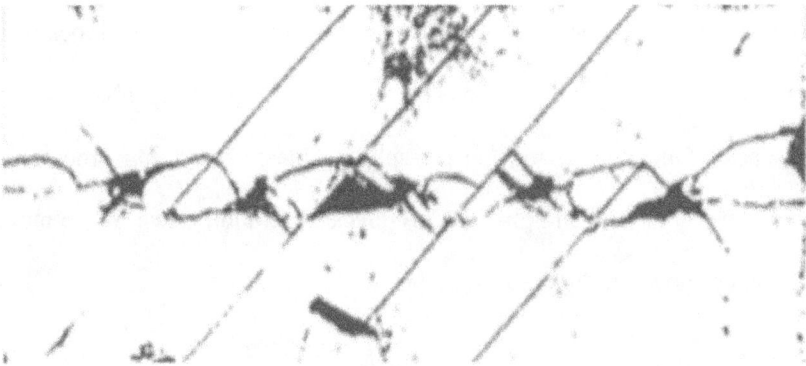

(a) α=45°

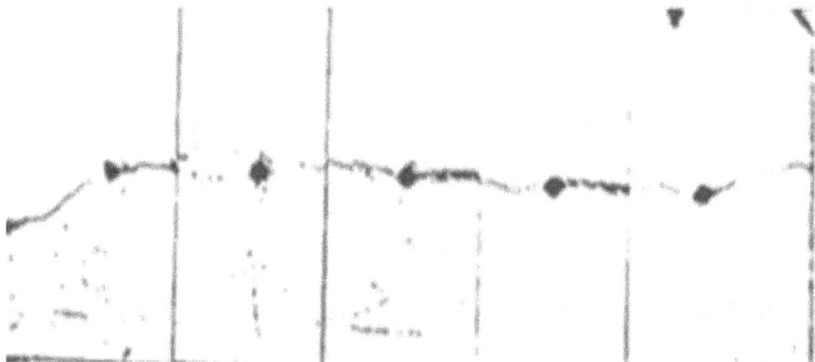

(b) α=90°

Fig.6.14 Modeling tests of pre-splitting blasting at two angles

The simulation results indicate that the forming effect under joint occurrence conditions at different angles is basically consistent with the theoretical analysis. These findings also align with the experimental results of jointed rock mass models reported by Gao Wenxue [85] and Worsey [86]. Compared with the joint model with an angle of α=45°, the jointed rock mass model with an angle of α=90° has better pre-splitting forming effect, and the pre-cracks generated during the pre splitting blasting forming process are more penetrating and straight. The pre-crack generation effect of the model test can be shown in Figure 6.14.

6.3 Engineering Verification

The Section ZK223+730-ZK223+970 of Longhua Expressway 26 is located near Xianqiao Underground River Scenic Area in Hengshitang Town, northwest of Yingde City, Guangdong Province. The terrain along this section is complex and varied. On the south side of this section, the nearest residential building is about 184m away from the slope crest and a 35KV high-voltage pole is about 100m away from the route. On the north side, the nearest residential building is about 196m away from the slope crest and a 110KV high-voltage pole is about 370m away from the route. The surrounding environment of the detonating area is relatively complex, as specifically shown in Figure 6.15.

Fig. 6.15 Surrounding environment of detonating area

The road cut in this section belongs to the geomorphological unit of the Beijiang Basin, which is characterized by erosion-accumulation landforms. The cutting area crosses an approximately northeast-southwest ridge. The existing terrain is steep, with a distribution elevation of 47-168m and a relative height difference of 121m. The longitudinal terrain slope along the route is 5°-

36°, with some areas showing steep ridges. The transverse terrain slope is steep, generally ranging from 15°-40°.

The strata in the blasting area are mainly composed of secondary red clay interbedded with block stones caused by the Quaternary Holocene landslide deposit, and the bedrock is composed of Devonian Upper Series Tianziling Formation limestone. The limestone in this section is gray in color, with a microcrystalline structure and medium to thick layered composition. It is mainly composed of carbonate minerals and locally filled with calcite. The moderately weathered zone exhibits relatively intact rock core, which is in the form of blocks and columns and has hard rock quality. Widely distributed in the cutting area, limestone is the main lithology of the cutting area.

The cutting area is located between the F1 and F2 fault in geological structure. Through geological mapping, the rock strata occurrence in the road cutting area is measured and the joint occurrence of the rock strata is measured to be 93-107°∠ 67-88°. No mudstone interlayer is found on the surface, indicating a hard structural surface with poor bonding. The strata joint and fissure occurrence in the cutting area vary greatly.

According to the lithological combination and groundwater occurrence conditions in the area, the groundwater types in the cutting area can be divided into two categories: Quaternary loose rock pore groundwater and bedrock karst water.

In summary, the terrain of the blasting slope in the blasting area is relatively steep, and the rock mass of the slope is mainly composed of strongly weathered limestone with hard rock quality and has developed joints and fissures. The geological conditions of the slope are good.

The slope of the blasting area is primarily composed of moderately weathered limestone. The northern slope consists of eight levels extending to the crest and the southern slope consists of six levels extending to the crest. The limestone slope is characterized by high hardness and stability. A wide platform separation is used for blasting excavation to form a natural slope. The slope ratio of each level is controlled at 1: 0.75, with varying widths for each level of the platform. The blasting area features a large height difference, reaching a maximum of 121m, thus falling into the category of high slope excavation. During blasting excavation, control blasting techniques such as smooth blasting or pre-splitting blasting should be used to excavate the contour line of the slope. Considering that the rock excavated by blasting is mainly used for filling in other road sections, there are specific requirements for the fragmentation of the blasted rock. Therefore, the grading requirements for rock blasting fragmentation should be taken into account in the design of blasting parameters, detonation network, and charge structure.

According to the specific on-site conditions of pre-splitting blasting construction on, the blasting area varies with each operation. Generally, the seven measuring points are arranged approximately 20-200m along the recoil direction from the blasting source area. Vibration measurement is mainly conducted for each measuring point in the blasting area, where vertical and horizontal radial velocity sensors are deployed to monitor the blasting excavation and analyze the vibration characteristics of the rock mass in the blasting area. The general arrangement of vibration measuring points for on-site pre-splitting blasting is shown in Figure 6.16.

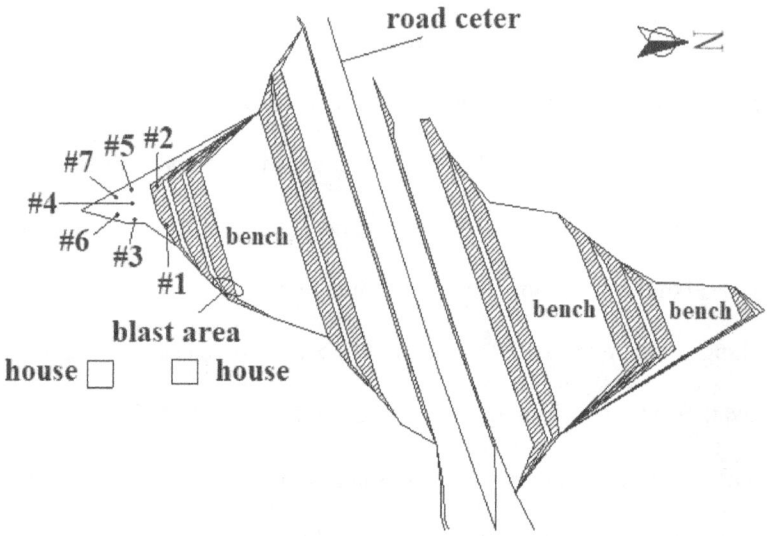

Fig.6.16 Layout of measuring points

By monitoring the blasting vibration, the vibration velocity of measuring points at different time intervals can be obtained and then through blasting experience and theoretical calculations, the stability of the slope can be determined. Therefore, real-time monitoring of slope blasting can be carried out. By collecting relevant measurement data such as blasting parameters, blasting center distance, and height difference, and conducting feature analysis of the monitored blasting vibration signals, the propagation law and hazard mechanism of blasting seismic waves can be investigated and the blasting parameters and blasting sequence can be reasonably determined so as to ensure the safety and stability of the slope. The blasting test system for this study consists of a blasting vibration meter, a vibration sensor, a PC machine, and a report printer, as shown in Figure.6.17.

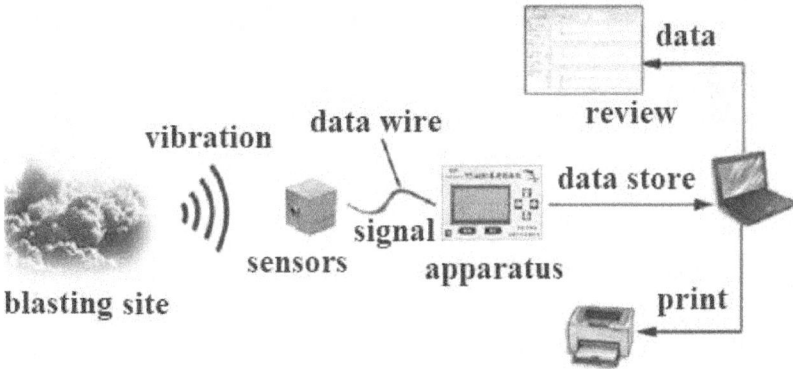

Fig.6.17 Blasting testing system

The main vibration measuring instrument used in this blasting monitoring system is the TC-4850 blasting vibration meter. Developed and produced by Chengdu Zhongke Measurement and Control Co., Ltd., this vibration measuring instrument is a portable vibration monitoring instrument dedicated to engineering blasting vibration measurement specially designed for production and research. With such advantages as small size, light weight, high reliability, simplicity and ease of use, and pressure resistance, this type of monitoring instrument can be matched with dedicated sensors to excellently complete the corresponding monitoring tasks. The layout of monitoring points for on-site pre-splitting blasting is shown in Figure 6.18.

Fig.6.18 On-site vibrometers arranged to monitor measuring points

Based on the specific on-site construction situations, Vibration monitoring tests are conducted on multiple pre-splitting blasting in the blasting area. The TC-4850 blasting vibration meter is used to obtain the maximum vibration velocity, main vibration frequency, maximum vector superposition vibration velocity, closest horizontal distance from the blasting source, and height difference in the X, Y, and Z directions of the measuring points #1, #2, #3, #4, #5, #6 and #7, as shown in Table 6.5.

Tab.6.5 Monitoring results of on-site pre-splitting blasting

No.	Pt.	charge (kg)	Dis. (m)		Peak velocity of particle (cm/S)				Main frequency (Hz)		
			Hor.	Ver.	X	Y	Z	sum	X	Y	Z
I	1	4992	68.9	17.7	3.36	6.68	6.87	10.15	15.18	12.66	17.7
I	2	4992	69.7	17.7	6.03	8.78	7.24	12.88	12.2	14.3	21.3
I	3	4992	113.2	25.3	3.03	3.14	3.95	5.89	35.2	10	28.2
I	4	4992	113.1	25.3	6.89	3.88	5.69	9.74	22.6	19.1	22.6

244

I	5	4992	113.7	25.3	4.72	5.1	5.80	9.05	20.62	40.67	37.44
I	6	4992	120.7	27.8	6.88	3.99	5.70	9.78	47.62	14.87	36.22
I	7	4992	120.8	27.8	4.15	5.55	7.06	9.89	22.73	13.56	35.09
II	1	1680	289.9	6.1	0.38	0.38	0.26	0.60	28.78	16.19	25.81
II	2	1680	271.9	6.1	0.87	0.83	0.57	1.33	32	22.86	31.5
II	3	1680	254.9	6.1	1.11	1.28	0.78	1.87	30.77	20	34.78
II	4	1680	230.9	7.3	0.42	0.61	0.50	0.89	12.74	20.94	30.08
II	5	1680	215.9	7.3	0.79	1.05	0.80	1.54	13.5	13.5	33
II	6	1680	214.2	18.5	0.28	1.72	1.28	2.16	16.5	15.2	2.3
II	7	1680	205.2	18.5	2.23	1.86	1.34	3.20	39.1	17.4	26.5
III	1	4998	310.0	0	0.34	0.21	0.33	0.52	18.6	12.35	14.65
III	2	4998	291.0	0	0.9	0.8	0.62	1.35	21.28	20.51	23.95
III	3	4998	279.0	0	0.72	0.88	0.62	1.30	19.61	19.23	23.95
III	4	4998	243.7	12.9	0.44	0.32	0.3	0.62	17.35	17.43	23.32
III	5	4998	237.7	12.9	0.78	0.47	0.6	1.09	13	13.5	18.7
III	6	4998	212.9	21.6	1.63	0.2	0.92	1.88	13.9	17.4	26.9
III	7	4998	218.9	21.6	1.89	0.94	1.57	2.63	27.4	17.8	26.9
IV	1	4080	239.3	17.9	0.51	0.48	0.71	1.00	14.87	17.94	74.07
IV	2	4080	219.3	17.9	1.74	1.33	0.73	2.31	22.73	31.25	23.67
IV	3	4080	207.2	17.9	1.19	1.65	1.01	2.27	21.74	28.37	30.08
IV	4	4080	160.9	4.7	1.11	0.97	1.57	2.15	40.02	40.77	41.03
IV	5	4080	177.9	4.7	0.6	1.41	1.1	1.89	13.9	45.6	36.9
IV	6	4080	151.9	5.2	2.89	1.6	1.75	3.74	20.4	20	20
IV	7	4080	161.9	~5.2	2.51	1.9	1.56	3.51	19.5	20.8	26.9
V	1	4080	72.2	17.1	5.61	7.8	5.6	11.12	57.14	46.51	35.09
V	2	4080	71.1	17.1	19.05	7.53	19.57	28.33	133.33	133.33	181.82
V	3	4080	72.3	17.1	10.09	10.98	13.32	19.99	8.99	9.9	50
V	4	4080	66.3	13.5	4.85	8.74	9.9	14.07	13.05	38.28	34.63

V	5	4080	90.8	4	6.16	5.22	7.32	10.90	23.9	16.1	15.2
V	6	4080	93.8	4	4.13	5.59	8.4	10.90	10.9	13.5	28.2
VI	1	3600	108.7	16.7	2.76	4.06	2.85	5.68	74.07	44.44	48.19
VI	2	3600	118.8	16.7	2.18	4.89	3.2	6.24	34.48	40.4	36.7
VI	3	3600	128.6	19.2	1.67	3.45	1.47	4.11	10.61	14.6	16.33
VI	4	3600	117.1	14.2	8.22	4.08	2.69	9.56	242.42	37.04	36.7
VI	5	3600	90.7	15.2	6.46	4.31	4.57	9.01	31.3	34.7	33
VI	6	3600	102.8	6.3	3.27	4.85	4.71	7.51	21.3	20.4	21.7
VI	7	3600	90.8	6.3	3.03	2.78	3.07	5.13	24.8	19.5	20

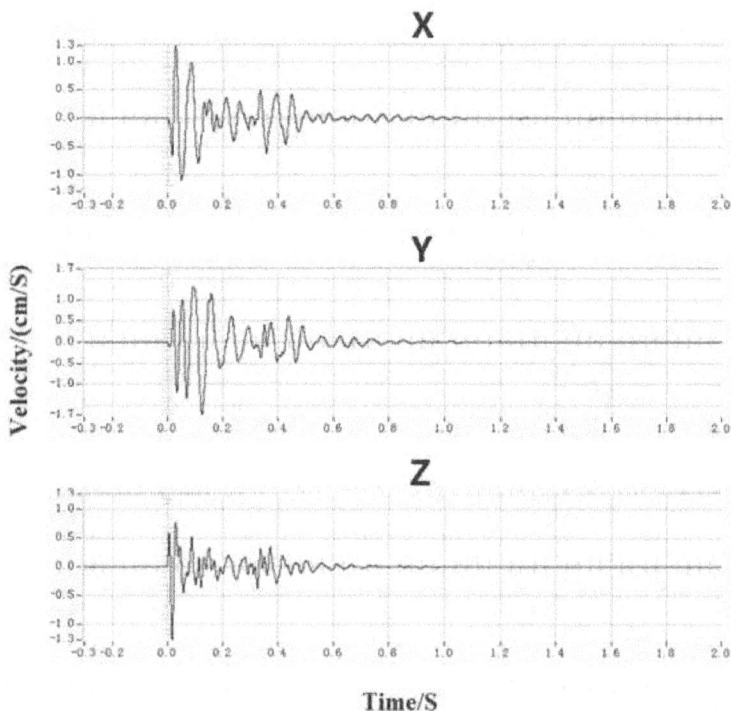

Fig.6.19 Pre-splitting blasting vibration waveform of 6#

Based on the actual situation of the construction site, the vibration

waveform of the second pre-splitting blasting monitoring point 6 # is selected as a typical monitoring point for analysis. The monitoring point is conducted at 11:33:51 on December 9, 2016. The vibration waveform of the particle blasting is shown in Figure 6.19.

According to the actual situations at the engineering construction site, the multiple pre-splitting blasting operations analyzed in this study exhibit distinct common characteristics: a relatively large total explosive charge. Due to the limited number of detonators used on site, coupled with the insufficient segmentation of the pre-splitting blasting process, which is manifested by the presence of only one peak, the operations can be categorized as a one-time detonation method.

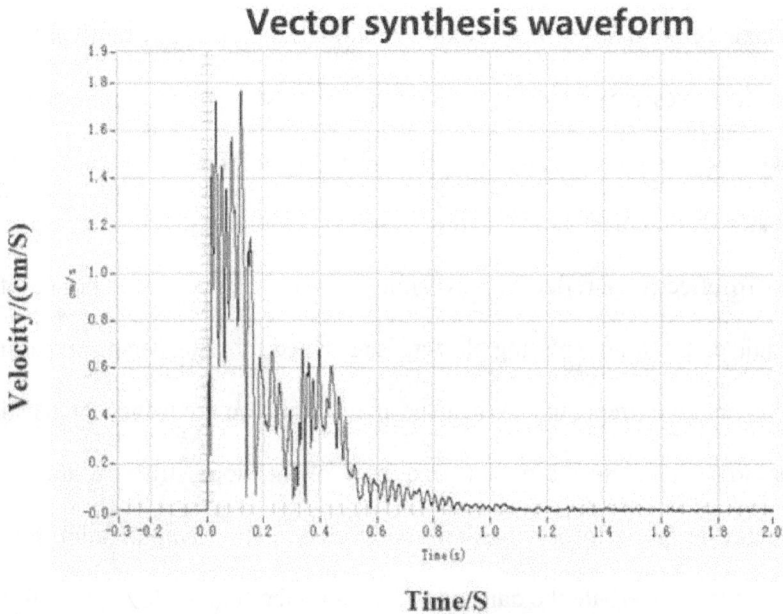

Fig.6.20 Pre-splitting blasting vibration vector synthetic waveform of 6#

As shown in Figure 6.20, the horizontal axis of the blasting vector synthetic vibration waveform diagram represents the blasting time, and the vertical axis represents the value of the synthetic vibration wave velocity. The diagram is consistent with the vibration waveform monitored on site, both with the presence of only one peak. It can be seen that the pre-cracks generated by the pre-splitting blasting have a shock-absorbing effect. Because the charge of the pre-splitting hole is smaller than that of the main blasting hole, the vibration velocity caused by the pre-splitting hole is smaller than that caused by the main blasting hole. The above waveform diagram shows that the peak of the blasting vibration velocity appears at a relatively earlier position, which can be considered as caused by the joint blasting of the pre-splitting hole and the main blasting hole. The blasting vibration decreases significantly after the seismic waves generated by the subsequent blasting pass through the pre-cracks, which also indicates that the weak structural surfaces such as pre-cracks and joint cracks jointly reduce the vibration hazard.

To effectively reflect the dynamic response of slope rock masses, the seismic effect of pre-splitting blasting is analyzed from the perspective of the main vibration frequency. When the main vibration frequency of blasting vibration is close to the natural frequency of the slope structure, the slope structure is prone to co-vibration and slope vibration amplification effect, which will exacerbate the damage of blasting vibration to slope pre-splitting formation. Therefore, in the study of the influence of blasting vibration on slope rock masses under joint occurrence conditions, the influence of the main vibration frequency of blasting vibration cannot be ignored. The statistics of the main vibration frequency of blasting vibration are shown in Table 6.6.

Tab.6.6 Statistics of main vibration frequency of pre-splitting blasting

Axis	No. of occurrence of main frequency (Times)											
	<10 Hz	15 Hz	20 Hz	25 Hz	30 Hz	35 Hz	40 Hz	45 Hz	50 Hz	55 Hz	60 Hz	>60 Hz
X	1	9	6	10	2	4	3	1	1	0	1	3
Y	2	4	13	7	1	2	2	4	2	0	0	1
Z	1	1	6	9	9	9	7	1	2	1	0	2

Table 6.6 presents the statistical data of the main vibration frequencies based on on-site blasting vibration monitoring. By statistically analyzing the main vibration frequencies of each vibration data with each interval being 5Hz, the main frequency distribution trends in the X, Y, and Z directions can be obtained, as shown in Figure 6.21.

From Figure 6.23, it can be clearly observed that there are significant differences in the distribution of the main vibration frequencies in the X, Y, and Z directions. The main vibration frequencies in the X and Y directions are mainly distributed within the range of 10-25Hz, while the main vibration frequencies in the Z direction are mainly distributed within the range of 15-40Hz.

Relevant data indicate that the natural frequency of jointed rock mass slopes is relatively low, generally concentrated between 2-10Hz. Therefore, the seismic wave frequency at the blasting vibration monitoring location is significantly close to the natural frequency of the jointed rock mass, indicating that the main vibration frequency of blasting vibration may have a certain impact on the structure of the slope rock mass. Further in-depth research and evaluation of the impact of blasting seismic waves on jointed rock mass slopes and their elevation effects are needed, and attention should be paid to the changes in blasting seismic wave frequency.

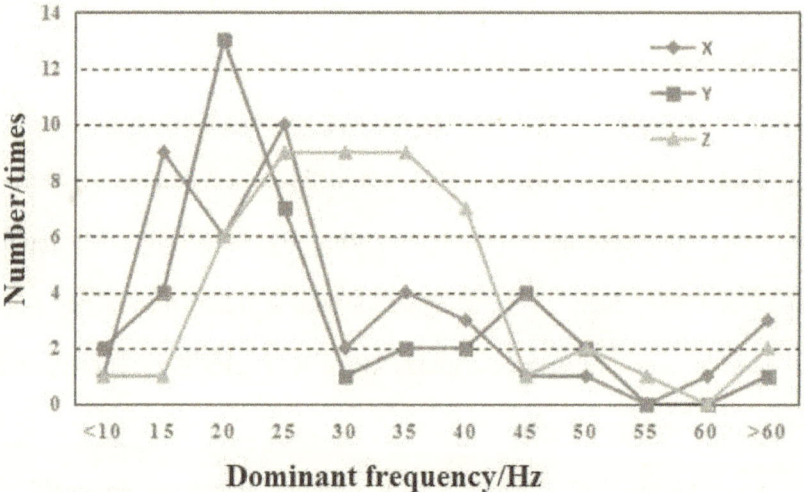

Fig.6.21 Distribution of main vibration frequencies in pre-splitting blasting

1. Regression formula without considering elevation effect

Reference [86] establishes a mathematical model using the Sadovsky empirical attenuation formula to predict the propagation and attenuation laws of blasting vibrations

$$V = K\left(\frac{\sqrt[3]{Q}}{R}\right)^{\alpha} \tag{6.17}$$

In the formula, V is the peak vibration velocity of the particle, cm/S; K, α is the geological condition factor and attenuation index; Q is the maximum explosive charge corresponding to the peak vibration velocity; R is the straight-line distance between the blasting source and the measuring point, m.

Based on the monitoring data obtained from the pre-splitting blasting site, the function equations of the attenuation propagation law of blasting vibration velocity in the X, Y, Z directions and vector superposition are analyzed and fitted, as shown in Figures 6.22 to 6.25.

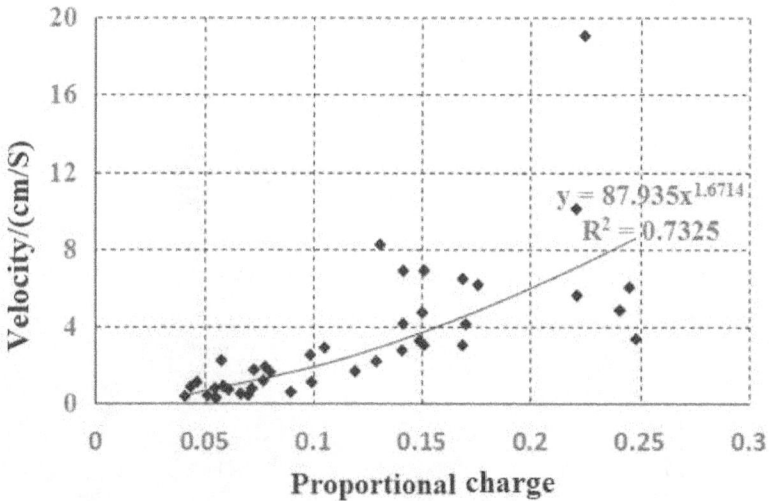

Fig.6.22 Fitting curve of blasting vibration velocity in X direction

Formula for velocity function of blasting vibration in X direction:

$$V_x = 87.935 \left(\frac{\sqrt[3]{Q}}{R} \right)^{1.6714} \tag{6.18}$$

Formula for velocity function of blasting vibration in Y direction:

$$V_y = 99.306 \left(\frac{\sqrt[3]{Q}}{R} \right)^{1.7264} \tag{6.19}$$

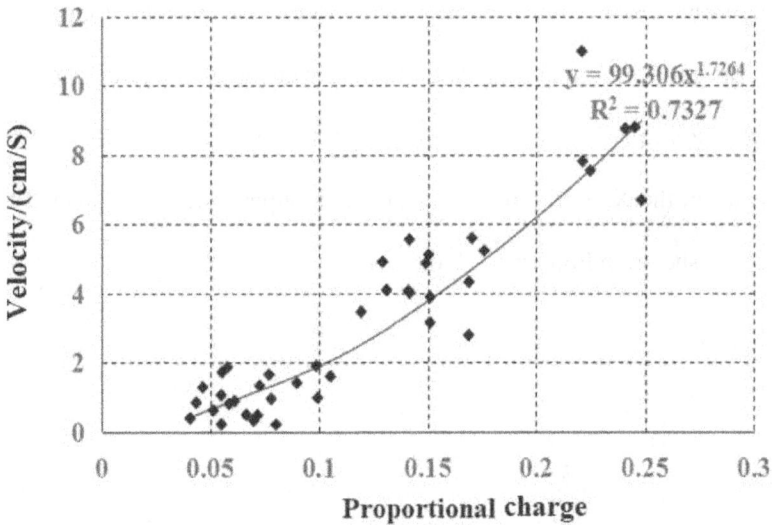

Fig.6.23 Fitting curve of blasting vibration velocity in Y direction

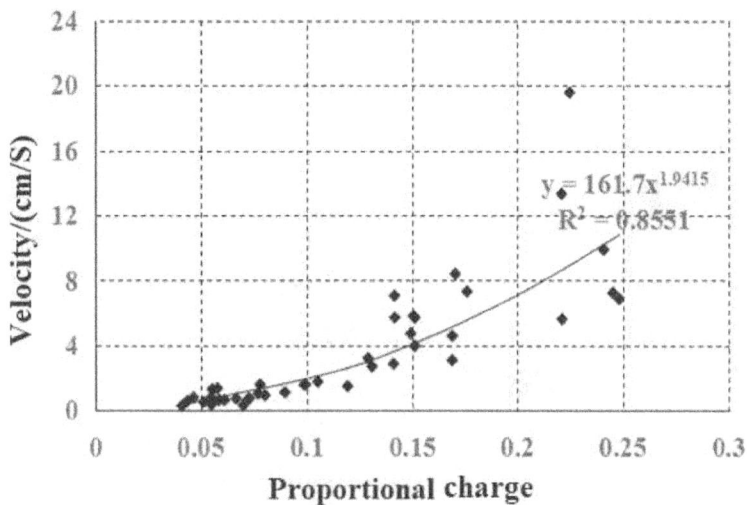

Fig.6.24 Fitting curve of blasting vibration velocity in Z direction

Formula for Z direction blasting vibration velocity function:

$$V_z = 116.7 \left(\frac{\sqrt[3]{Q}}{R} \right)^{1.9415} \qquad (6.20)$$

Formula for velocity function of vector superposition blasting vibration:

$$V = 202.2 \left(\frac{\sqrt[3]{Q}}{R} \right)^{1.7656} \qquad (6.21)$$

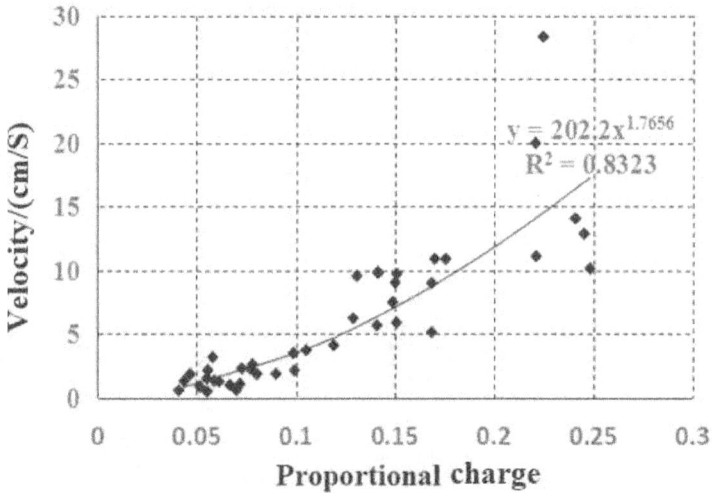

Fig. 6.25 Fitting curve of blasting vibration velocity in vector superposition

2. Regression formula considering elevation effect

Equation (6.17) is applicable to flat terrain and does not take into account the influence of slope elevation on the propagation law of blasting vibration velocity. For slopes with large height differences, the elevation amplification effect of the slope should also be considered. Therefore, the empirical formula [87] can be referred to:

$$V = K\left(\frac{\sqrt[3]{Q}}{D}\right)^{\alpha}\left(\frac{\sqrt[3]{Q}}{H}\right)^{\beta} \tag{6.22}$$

In the formula, V is the peak vibration velocity of the particle, cm/S; K, α, β represent geological conditions and attenuation index; Q is the maximum explosive charge segment corresponding to the peak vibration velocity; R is the straight-line distance between the blasting source and the measuring point, m; H is the straight-line distance between the explosive source and the measuring point, m.

For the monitoring data measured at the pre-splitting blasting site, the analysis and fitting of the function equations for the attenuation and propagation laws of blasting vibration velocity in the X, Y, Z directions and after vector superposition are conducted respectively.

The function formula for the blasting vibration velocity in the X direction is

$$V_x = 60.635 \left(\frac{\sqrt[3]{Q}}{R} \right)^{1.4259} \left(\frac{\sqrt[3]{Q}}{H} \right)^{0.4248} \tag{6.23}$$

The function formula for the blasting vibration velocity in the Y direction is

$$V_y = 82.296 \left(\frac{\sqrt[3]{Q}}{R} \right)^{1.5869} \left(\frac{\sqrt[3]{Q}}{H} \right)^{0.0831} \tag{6.24}$$

The function formula for the blasting vibration velocity in the Z direction is

$$V_z = 157.625 \left(\frac{\sqrt[3]{Q}}{R} \right)^{1.8209} \left(\frac{\sqrt[3]{Q}}{H} \right)^{0.1663} \tag{6.25}$$

The function formula for vector superposition velocity is

$$V = 157.625 \left(\frac{\sqrt[3]{Q}}{R} \right)^{1.6022} \left(\frac{\sqrt[3]{Q}}{H} \right)^{0.1322} \tag{6.26}$$

In conducting regression analysis of the pre-splitting blasting vibration velocity equation, the correlation coefficient can serve as an important measurement indicator. It is generally believed that the larger the correlation coefficient of the equation is, the smaller the equation error is. Therefore, as the correlation coefficient becomes larger, the credibility of the equation also becomes higher. A comparative analysis of the correlation coefficients of the blasting vibration velocity equations in various directions without considering the elevation effect and with considering the elevation effect is presented in Table 6.7.

Tab. 6.7 Correlation coefficient of the vibration velocity of pre-splitting blasting

Direction	Not considering elevation effects	Considering elevation effects
X	$R_2=0.7325$	$R_2=0.4248$
Y	$R_2=0.7327$	$R_2=0.4552$
Z	$R_2=0.8551$	$R_2=0.6514$
Vector superposition	$R_2=0.8323$	$R_2=0.6853$

As can be seen from Table 6.7, when comparing the correlation coefficients of the equations obtained by fitting the pre-splitting blasting vibration velocities using equations (6.18) and (6.21), the pre-splitting blasting vibration velocity fitting equations without considering the slope elevation effect have a higher degree of correlation than those considering the elevation effect. Equation (6.17) has a higher regression accuracy and therefore it is recommended that this equation be prioritized for regression calculation in this area. However, the regression accuracy of equation (6.22) is relatively low, which may be due to the blocking effect of joints and fractures, resulting in the absence of the elevation amplification effect. Further in-depth research is still required.

In addition, it can be found that the correlation coefficient of the blasting vibration velocity equation in the Z direction without considering the elevation effect is the largest, indicating that the empirical formula without considering the elevation effect can accurately and effectively describe the propagation law of vibration velocity in the Z direction during the pre-splitting blasting of rock masses with joint occurrence conditions.

This chapter conducts a theoretical analysis of the reflection and transmission phenomena of blast-induced seismic waves in jointed rock masses. Combined with the actual engineering geological conditions of the blasting area in TJ26 section of Longhua Expressway, the pre-splitting blasting of the cutting slope in this section is monitored, and the results of vibration monitoring are analyzed and studied.

(1) The propagation and attenuation laws of blasting seismic waves are closely related to the topographic conditions of the blasting area. The presence of joints affects the propagation of seismic waves in rock masses. When seismic waves reach weak structural surfaces such as joints, reflection and transmission phenomena will occur, which will have a certain blocking effect on the propagation of seismic waves.

(2) Based on the actual engineering geological conditions of the blasting

area in TJ26 section of Longhua Expressway, monitoring points are reasonably selected. TC-4850 blasting vibrometer is used to monitor the pre-splitting blasting of the rock cutting in the blasting area, and data such as vibration velocity and main frequencies are measured.

(3) To simulate the influence of joints at different angles on the blasting effect, six typical jointed rock mass models with angles α of 0°, 30°, 45°, 60°, 75°, and 90° between the joint and the line connecting boreholes are selected for numerical simulation of pre-splitting formation. The results show that when the angle α is 45°, the pre-splitting effect is the poorest, and overbreak or underbreak is likely to occur during the slope pre-splitting process. When the angle α is 90°, the presplitting effect is the best, as the joint has minimal impact on the pre-splitting outcome, and the post-blast contour line largely matches the experimental results of the blasting test.

References

[1] Cui Zuozhou, Yin Gexun, Gao Enyuan, et al. *Comprehensive Study on the Velocity Structure and Deep Structure of the Qinghai Tibet Plateau in the Yadong Golmud Rock Circle Geological Section*[M]. Beijing: Geological Publishing House, 1992.

[2] Gao Dazhao. Review and Prospect of Geotechnical Engineering[M]. Beijing: People's Communications Press, 2001.

[3] Xu Xiao. *Study on Dynamic Response of Mudded Intercalations during Cyclic Loading*[D]. Zhengzhou: Zhengzhou University, 2019.

[4] Tang Liangqin, Liu Dongyan, Nie Dexin. Strength Parameter Selection of Weak Intercalated Layer in Xiangjiaba Bed Foundation[J]. *Journal of Zhejiang University (Engineering Science Edition)*, 2013, 47(1): 162-169.

[5] Qu Yongxin. From the Bentonite Incident at the Chushandian Reservoir Project to Prediction Research of Claystone Mudzation at the Gezhouba Project[J]. Journal of Engineering Geology, 2014, 22(04): 699-702.

[6] Zheng Wentang, Xu Weiya, Zhang Zhiliang, et al. Stability of High Rock Slope Formed by Complex Wedge[J]. Journal of Yangtze River Academy of Sciences, 2008, 25(5): 115-119.

[7] Xie Yushou, Wang Yaowen. Seismic Effects of Industrial Blasting[J].

Chinese Journal of Geophysics, 1962(02): 154-163

[8] Zhang Jietao, Wang Jianzhong, Guo Xuebin. Measurement and Analysis of Blasting Particle Vibration in Adjacent Area of Zone[J]. Journal of Safety and Environment, 2003(02): 36-39.

[9] Sun Xiumin, Wang Jinyu. Evaluation of Blasting Vibration Effects on Buildings[J]. Blasting, 2008(02): 73-76+88.

[10] He Yunlong. An Approximate Calculation Method for the Accelerations in the Rock Slope under Blasting Construction[J]. Journal of Rock Mechanics and Engineering, 1996(01): 19-25.

[11] Hu Guozhong. Forecast on Intensity of Vibration and Study on Effect of Ground Blast Induced Vibration of Underground Engineering in City[D]. Chongqing: Chongqing University, 2005.

[12] Zhang Yongzhe. Study on Characteristics of Blasting-Caused Seismic Wave[J]. Blasting, 2000(S1): 6-10.

[13] Pang Huandong, Chen Shihai. Variation Law of Blasting Seismic Wave's Propagation in Elastic Media[J]. Vibration and Shock, 2009, 28(03): 105-107+202.

[14] Wu Delun, Ye Xiaoming. A Comprehensive Review and Commendation of Blast Vibration Safety Velocity[J]. Journal of Rock Mechanics and Engineering, 1997(03): 67-74.

[15] Tang Chunhai, Yu Yalun, Wang Jianzhou. Elementary Study of Safety Criterion for Blasting Vibration[J]. Nonferrous Metals, 2001(01): 1-4.

[16] Zhang Tianjun, Ma Rui, Qiao Baoming, et al. The Regression Analysis

of Extended Sadovsky Type in Blasting Vibration[J]. Journal of Hunan University of Science and Technology (Natural Science Edition), 2012, 27(01): 12-16.

[17] Yang Nianhua, Zhang Le. Blasting Vibration Waveform Prediction Method Based on Superposition Principle[J]. Explosion and Shock, 2012, 32(01): 84-90.

[18] Hao Quanming, Zhang Zheng, Chang Jianping. Free Surface Azimuth of Blasting Vibration Velocity in Drilling Blasting[J]. Coal Technology, 2015, 34(02): 328-330.

[19] Tao Tiejun, Wang Xuguang, Chi En'an, et al. Attenuation Formula of Blasting Seismic Wave Based on Energy Theory[J]. Engineering Blasting, 2015, 21(06): 78-83.

[20] Wang Yujie, Liang Chaoshui, Tian Xinbang. Study on Redundant Regulation of Underground Digging Blasting Vibration of Zhouning Hydropower Station[J]. Chinese Journal of Rock Mechanics and Engineering, 2005(22): 4111-4114.

[21] Zhang Shaoquan, Guo Jianming. On the Coupling Factor of an Explosion[J]. Acta Geophysica Sinica, 1984(06): 537-547.

[22] Tang Hai, Li Haibo. Study of Blasting Vibration Formula of Reflecting Amplification Effect on Elevation[J]. Rock and Soil Mechanics, 2011, 32(03): 820-824.

[23] Lou Jianwu, Long Yuan, Xu Quanjun, et al. A Study on the Extraction and Prediction of Blasting Seismic Wave Characteristics by Wavelet

Packets Technique[J]. Explosion and Shock, 2004(03): 261-267.

[24] Ling Tonghua, Li Xibing. Analysis of Energy Distributions of Millisecond Blast Vibration Signals Using the Wavelet Packet Method[J]. Chinese Journal of Rock Mechanics and Engineering, 2005(07): 1117-1122.

[25] Li Hongtao, Lu Wenbo, Shu Daqiang, et al. Study of Energy Attenuation Law of Blast-Induced Seismic Wave[J]. Chinese Journal of Rock Mechanics and Engineering, 2010, 29(S1): 3364-3369.

[26] Zhou Junru, Lu Wenbo, Zhang Le, et al. Attenuation of Vibration Frequency during Propagation of Blasting Seismic Wave[J]. Chinese Journal of Rock Mechanics and Engineering, 2014, 33(11): 2171-2178.

[27] Tang Hai, Li Junru. Numerical Simulation of Influence of Protruding Topography on Blasting Vibration Wave Propagation[J]. Rock and Soil Mechanics, 2010, 31(04): 1289-1294.

[28] Yu Min, Lin Congmou, Chang Fangqiang, et al. Research on Amplification Effect of Blasting Vibration in Rock Deep Foundation Pit[J]. Blasting, 2017, 34(04): 27-32+65.

[29] Cao Pan, Yan Shilong, Ni Lei, et al. Research on Attenuation Law of Explosion Stress Wave in Rock by UDEC Modeling[J]. Blasting, 2014, 31(01): 42-46.

[30] Ye Haiwang, Zhou Jianmin, Yu Hongbing, et al. Influence of Rock Structure Planes on Propagation of Blasting Seismic Wave[J]. Blasting, 2016, 33(01): 12-18.

[31] Jiang Nan, Zhou Chuanbo, Ping Wen, et al. Altitude Effect of Blasting Vibration Velocity in Rock Slopes[J]. Journal of Central South University (Natural Science Edition), 2014, 45(01): 237-243.

[32] Lu Wenbo, Zhang Le, Zhou Junru, et al. Theoretical Analysis on Decay Mechanism and Law of Blasting Vibration Frequency[J]. Blasting, 2013, 30(02): 1-6+11.

[33] Zhou Jianmin, Yue Menglei, Wang Xuguang, et al. The Numerical Simulation Study on the Influence of Rock Structure Planes on the Dynamic Stability of Slope[J]. Mining Research and Development, 2017, 37(08): 1-4.

[34] Li Hongtao. Study on Effect of Blast-induced Seismic Based on Energy Theory[D]. Wuhan: Wuhan University, 2007.

[35] Rockwell, E H. Quarry Blasting Vibrations and Their Effect on Structures[J]. Cement, Mill Quarry Engineering, 1927, 30(10): 30-34.

[36] Jenkins, J E. Scientific Method of Measuring Vibrations from Quary Blasting[J]. Pit Quarry, 1945, 38(02): 67-68.

[37] Leet, L D. Vibration from Blasting[J]. Explosives Engineering, 1946, 24(03): 85-89.

[38] Morris, W R. Damage to Structures by Ground Vibrations Due to Blasting[J]. Mine Quarry Eng., 1953, 19(04): 116-118.

[39] Carlos Lopez. Drilling and Blasting of Rocks[J]. Printed In Netherlands, 1995.

[40] Devine, J F. Effect of Charge Weight on Vibration Levels from Quarry

Blasting[M]. US Department of the Interior, Bureau of Mines,1966.

[41] Duvall, W I, Fogelson, D E. Review of Criteria for Estimating Damage to Residences from Blasting Vibrations. US Bureau Mines Report Investigations 5968,1962:1-19.

[42] Dowding, C H. Blast Vibration Monitoring and Control[M]. Englewood Cliffs: Prentice Hall, 1985.

[43] Zhang Yiping, Wu Guiyi. Study on Characteristics of Blasting-Caused Seismic Wave[J]. Mining Research and Development, 2007(06): 68-72.

[44] Zhang Yongzhe. Study on Propagation Characteristics of Explosive Seismic Wave[J]. Blasting, 2000(S1): 6-10.

[45] Li Aichen. Study on Dynamic Response and Stability of Open Rock Slope under Blasting Vibration[D]. Wuhan: Wuhan University of Science and Technology, 2020.

[46] Meng Qinghao, Ouyang Tianyun, Li Aichen. Numerical Simulation Analysis of Propagation Law of Blasting Seismic Waves[J]. Opencast Mining Technology, 2020, 35(04): 28-31.

[47] Lin Daneng, Hu Wei, Peng Gang. Analysis on Blasting Extrusion Characteristic of Soil and Rock[J]. Chinese Journal of Rock Mechanics and Engineering, 2003(11): 1767-1770.

[48] Long Yuan, Feng Changgen, Xu Quanjun, et al. Study on Propagation Characteristics of Blasting Seismic Waves in a Rock Medium and Numerical Calculation[J]. Engineering Blasting, 2000(03): 1-7.

[49] He Xiaoguang, Zhang Gansheng. Discussion on the Safety Criterion

from Detonated Seism[J]. Journal of Liaoning Institute of Science and Technology, 2005(02): 26-29.

[50] Zhang Zuyuan, Wang Hailiang, Li Lin, et al. Extension Technology of Shallow Tunneling for Civil Air Defense Work Close to Existing Building[J]. Track Traffic & Underground Engineering, 2014, 32(06): 74-78.

[51] Sun Cuiyuan, Xue Li, Meng Haili, et al. Experimental Research on Several Typical Blasting Vibration-Controlling Technologies[J]. Transactions of Beijing Institute of Technology, 2018, 38(04): 359-363+370.

[52] Huang Zheng, Song Yaodong, Zhang Huiping, et al. Vibration Control Technology for Open-pit Deep Hole Blasting[J]. Mining Technology, 2019, 19(05): 149-151.

[53] Li Yi, Teng Yong, Hu Kunlun, et al. Blasting Vibration Control Technology and Detection of Super High Buildings[J]. Initiators & Pyrotechnics, 2019(02): 46-49.

[54] Feng Na. Study on the Influence of Open-pit Blasting Vibration on Buried Oil Pipeline[D]. Qingdao: Shandong University of Science and Technology, 2020

[55] Zhang Guojun. Softening and Argillitization Processes of Sandwich and Chemical Kinetics Simulation[D]. Nanjing: Hohai University, 2007.

[56] Chen Zhixiong. ANN Method in Highway Tunnel Slope Design[D]. Chongqing: Chongqing University, 2005.

[57] Ren Jie. Study on Sliding Mode and Deformation Law of Bedding Rock Landslide with Double Weak Interlayers[D]. Taiyuan: Taiyuan University of Technology, 2019.

[58] Yao Zhi. Models and Sensitive Strata to Devolution and Landslip in Western Guizhou[J]. Guizhou Geology, 1994(03): 224-233.

[59] Wang Anxiang. The Influencing Factors and Zoning Evaluation of Landslides in Guizhou Province[J]. Environmental Science and Technology, 1995(01): 14-17.

[60] Wang Guirong. Research Status of Several Major Engineering Geological Problems Related to Weak Interlayers[J]. Water Resources and Hydropower Technology, 1987(11): 20-26.

[61] Hu Tao, Ren Guangming, Nie Dexin, et al. Strength Characteristics with Genesis and Type of the Sedimentary Weak Intercalated Layers[J]. Chinese Journal of Geological Hazards and Control, 2004(01): 127-131.

[62] Li Jingshan, Zhao Shanguo, Hu Ping. The Causes and Characteristics of Weak Interlayers in Dam Foundations[J]. Heilongjiang Science and Technology of Water Conservancy, 2007(02): 181.

[63] Ren Guangming, Nie Dexin. Discussion on the Strength Regeneration Characteristics and Mechanism of Soil Structure in Large Landslide Sliding Zone[J]. Hydrogeology and Engineering Geology, 1997(03): 28-31+44.

[64] Mahr, T. Deep Reaching Gravitational Deformation of High Mountain Slopes[J]. Engineering Geology, 1977, 16: 121-123.

[65] Tan T K, Li K R. Relaxation and Creep Properties of Thin Interbedded Clayey Seams and Their Fundamental Role in the Stability of Dams[C]. Proceedings of the International Symposium on Weak Rock. [S1]: A. A. Balkema, 1981: 369-374.

[66] Liu Xiaoli, Deng Jianhui, Li Guangtao. Shear Strength Properties of Slip Soils of Landslides: An Overview[J]. Rock and Soil Mechanics, 2004(11): 1849-1854.

[67] Jian Wenxing, Yin Kunlong, Ma Changqian, et al. Characteristics of Incompetent Beds n Jurassic Red Clastic Rocks in Wanzhou[J]. Rock and Soil Mechanics, 2005(06): 901-905+914.

[68] Lu Haifeng, Chen Congxin, Shen Qiang, et al. Genesis and Characteristic of Weak Interlayer Existed in the Red-bed Slope of Badong Formation in Southwestern Hubei Province[J]. Hydrogeology and Engineering Geology, 2010, 37(01): 54-61.

[69] Zeng Feng, Peng Jing. Study on Weak Interlayer in "Red Layer" Area[J]. Yangtze River, 2011, 42(22): 15-17.

[70] Huang Qiuxiang, Wang Jialin. Study of the Deformation Characteristics of an Anti-dip Slope with Soft Internal Layers[J]. China Civil Engineering Journal, 2011, 44(05): 109-114.

[71] Zhou Fei, Xu Qiang, Liu Hanxiang, et al. Shaking Table Model Test on Deformation and Failure Characteristics of Slope with Horizontal Weak Interlayer[J]. Mountain Research, 2014, 32(05): 587-594.

[72] Liu Hanxiang. Seismic Responses of Rock Slopes in a Shaking Table

Test[D]. Chengdu: Chengdu University of Technology, 2014.

[73] Song Yanqi, Li Ming, Liu Jiang, et al. Experimental Test on Marble Containing Natural Weak Interlayer of Different Angles[J]. Journal of China University of Mining and Technology, 2015, 44(04): 623-629.

[74] Liu Chuanzheng, Zhang Jianjing, Cui Peng. Energy Evolution and Stress Response during Stress Wave Prorogation in the Intercalation[J]. Rock and Soil Mechanics, 2018, 39(06): 2267-2277.

[75] Yu Jingtao, Yang Haifeng. Numerical Analysis of the Influence of Weak Interlayers with Different Inclination Angles on the Stability of Tunnel Surrounding Rock[J]. Highway Transportation Technology (Applied Technology Edition), 2013, 9(05): 194-196.

[76] Sun Jinshan, Liu Guiying, Zhou Xiaofei, et al. Attenuation Characteristics of Blasting Seismic Waves Passing through Weak Interlayer[J]. Engineering Blasting, 2017, 23(05): 5-8.

[77] Sun Jinshan, Li Zhengchuan, Liu Guiying, et al. Dynamic Stress and Vibration Characteristics of Geomaterials in Slopes Induced by Blasting Vibration[J]. Journal of Vibration and Shock, 2018, 37(10): 141-148.

[78] Huang Feng, Zhu Hehua, Xu Qianwei, et al. The Effect of Weak Interlayer on the Failure Pattern of Rock Mass around Tunnel-Scaled Model Tests and Numerical Analysis[J]. Tunnelling and Underground Space Technology, 2013, 35(APR.): 207-218.

[79] Du Ruifeng, Pei Xiangjun, Jia Jun, et al. Experimental Study on Rock Slope with Weak Interlayer under Blasting Seismic Wave[J]. China Civil

Engineering Journal, 2021, 54(04): 95-106.

[80] Pei Xiangjun, Cui Shenghua, Huang Runqiu. A Model of Initiation of Daguangbao Landslide: Dynamic Dilation and Water Hammer in Sliding Zone during Strong Seismic Shaking[J]. Chinese Journal of Rock Mechanics and Engineering, 2018, 37(02): 430-448.

[81] Cui Shenghua, Pei Xiangjun, Huang Runqiu. An Initiation Model of DGB Landslide: Non-coordinated Deformation Inducing Rock Damage in Sliding Zone during Strong Seismic Shaking[J]. Chinese Journal of Rock Mechanics and Engineering, 2019, 38(02): 237-253.

[82] Cui Shenghua, Pei Xiangjun, Huang Runqiu, et al. Excess Interstitial Water Pressure within Sliding Zone Induced by Strong Seismic Shaking: An Initiation Model of the Daguangbao Landslide[J]. Chinese Journal of Rock Mechanics and Engineering, 2020, 39(03): 522-539.

[83] Zhang Liming. Research on Effect of the Medium-length Hole Bench Blasting Vibration on Underground Pipelines[D]. Guizhou: Guizhou University, 2015.

[84] Wei Haixia. Study on Dynamic Response and Safety Criterion of Buildings to Blasting Vibration Waves[D]. Qingdao: Shandong University of Science and Technology, 2010.

[85] Gao Wenxue, Liu Qingrong. Study of the Pre-splitting Blasting in Rock Block[C]. Selected Papers from the Second National Conference on Rock Dynamics. Wuhan: Wuhan University of Surveying and Mapping Press, 1990.

[86] Worsey, P N, Farmer, I W, Matheson, G D. The Mechanics of Presplitting in Discontinuous Rock[C]. The 22nd U.S. Symposium on Rock Mechanics (USRMS). Cambridge: American Rock Mechanics Association, 1981.

[87] SL47-2020. Specifications of Excavation Blasting for Hydropower and Water Conservancy Project[S]. Electric Power Industry Standard of the People's Republic of China.

www.ingramcontent.com/pod-product-compliance
Lightning Source LLC
Chambersburg PA
CBHW021033210326
41598CB00016B/1000